Management for Engineers

Steven Henderson
Southampton Institute of Higher Education

Robert Illidge
Leicester School of Business, De Montfort University

Peter McHardy
Leicester School of Business, De Montfort University

BUTTERWORTH
HEINEMANN

Butterworth-Heinemann Ltd
Linacre House, Jordan Hill, Oxford OX2 8DP

ℛ A member of the Reed Elsevier plc group

OXFORD   LONDON   BOSTON
MUNICH   NEW DELHI   SINGAPORE   SYDNEY
TOKYO   TORONTO   WELLINGTON

First published 1994

British Library Cataloguing in Publication Data
Henderson, Steven
  Management for engineers
  I. Title   II. Illidge, Robert   III. McHardy, Peter
  658.002462

ISBN 0 7506 0673 8

Composition by Scribe Design, Gillingham, Kent
Printed and bound in Great Britain

MANAGEMENT FOR ENGINEERS

Contents

Figures

Case studies

Introduction

Practising engineers are designers, builders, planners, problem solvers, decision makers and, above all, innovators. But recent evidence suggests that their education and training concentrate on the designing and building roles. Beginning with a broad based science background, the young engineer is trained to a high level in a relatively narrow range field of technical expertise.

Once in practice, engineers quickly find that their contribution to the firm is limited if there has been no exposure to business and managerial processes. The modern engineer is required to co-operate with accountants, marketers, the salesforce and many others in order to apply their technical expertise to best effect, and offers crucial inputs to these other functions. In a few cases, a small business component of an engineering degree course or qualification may offer a limited guide to the business processes involved. In other cases, an economics course will simply mystify things further.

Consequently, there has been a growing trend in recent years for practising engineers to seek further training in business and management skills. However, it is the authors' contention that an appreciation of business and management skills should be gained during the progression towards professional status, particularly in the early stages of undergraduate and post-graduate courses.

Indeed, in the UK the need for engineers to have both engineering skills and commercial competence has never been so pressing. Business conditions of twenty years ago were often easier for engineers and engineering companies than today's fiercely competitive markets. In the past, demand for engineering products was often greater than the volume that sellers could produce and distribute. Consequently, engineering activity was commonly directed at producing ever greater outputs of standardized units. Smaller production runs of products customized for a particular buyer were often treated as an irritating side issue, managed poorly and manufactured inefficiently and expensively; unless the firm in question was a small, specialist supplier. Similarly, product innovation was not always regarded as a mainstream activity, and product perfection was sometimes held as more important that working to deadlines and budgets.

A modern engineer is no longer in a position to rely on the sales department to sell anything that is designed and produced for the convenience of the seller. Most engineering markets are now very competitive, with the industrial and commercial buyers in a position to insist on individual standards of design, performance, quality, features and after-sales service. Product innovation is no longer a secondary activity, but an activity vital for survival, carried out with full urgency against tight deadlines. In short, the main contribution of the engineer is to ensure competitiveness, by providing value for customers at the lowest cost possible.

This belated customer orientation requires the engineer's talent at the interface between marketing and engineering. Unfortunately many engineers have little formal training in the language and techniques used by the marketers to enter and dominate growing markets, exit from stagnant and declining markets and devise new methods providing value to customers. Failure to co-ordinate the marketing and engineering functions will lead to innovations which are not valued by buyers, missed deadlines and lost orders.

Where such integration has been successful, it has had profound implications for the manufacturing process. Production runs have become smaller, making investment in capital equipment much riskier, and manufacturing systems are required to be much more flexible. Moreover, buyers' demands drive quality targets ever higher. Yet tough competition often prevents the producer passing the costs of these improvements and innovations on to buyers in the form of higher prices. Consequently, design engineers must cater for economic production methods and contribute substantially to the control of costs when designing products to match a customer specification. Unless the processes are well managed, there will be occasions where design engineers, manufacturing engineers, marketers and accountants work at cross purposes, to the detriment of the whole organization.

Industrialists necessarily focus on satisfying customers and planning their operations strategically. In an ideal world, the engineer would be well placed to exercise a crucial role, by producing products which satisfy the needs of the buyer using the most technically efficient and cost conscious methods. Unfortunately, the opportunity is often lost and the engineer's vital contribution wasted, as the engineer is continually constrained by the conflicting needs and objectives of other departments within the same firm. In this position, the engineer is merely well placed to receive blame and criticism for mistakes.

To some degree, acceptance of the blame may be self-reinforcing. First, many engineers receive little formal education in accounting and marketing. But more importantly engineering talent is wasted outrageously by the rigid organizational structure still found in many manufacturing organizations. In addition to a sound grounding in marketing, strategy, finance and production, this book will introduce the reader to the idea of project teams, as an alternative means of deploying engineering talent within the context of change management.

Most of the material in the book was piloted on the 1991–92 and 1992–93 intakes of students on the industrial and business systems; physics with business; and chemistry with business BSc, degrees at DeMontfort University. Our thanks to them for their patience and helpful criticism. Similar thanks go to their tutors, Angela Zvesper, Domonic Elliott, Karin Munton, Ian Christison and Ray Rue. We also wish to express gratitude to Zena Cumberpatch for her work on the Notts Knitting case study, and Angela Zvesper for her editing and sanitizing of the case study Funfurs. Dr Scott receives our special thanks for allowing us to use his analysis of Pilkington plc. Amanda Giddins and Richard Phillips of Xylogics are thanked for their help, frankness and support.

Finally, since the authors have all contributed to every part of the book, our names appear in alphabetical order.

Steven Henderson
Robert Illidge
Peter McHardy

1 Engineers in an organizational context

The organizational problem

Engineers are designers, builders, planners, problem solvers, scientists and decision makers, but, above all, innovators. They apply their skills to finding innovative solutions to all kinds of mechanical, chemical, electrical, structural, construction, software systems and robotics problems. The fruits of their innovations are found across a whole range of industries, including manufacturing, printing, publishing, communications, utilities, retailing, distribution and many other sectors. Many engineers are self-employed or consultants, while many more work for large-scale multinational companies, small jobbing contractors or for other enterprises seeking to trade competitively. A large number are employed in such newly privatized industries as gas, water and electricity, where the rigours imposed by competition and business considerations are relatively new, while technical people in the NHS, local government and the remaining state sector, are just beginning to feel the disciplines imposed by the need to innovate within modern business constraints.

However, this innovation does not occur in a vacuum. The engineer will normally work within a team, and, more widely, will be part of the organization in which the innovative ideas will have their impact. Even in relatively small organizations, marketing, accounting, manufacturing, personnel, service provision and innovation activities will be carried out by different groups of people. They are often organized into separate, though overlapping, departments, each department applying a range of special techniques, technologies and managerial skills to tackling different aspects of the same business problem. As an organization grows in size, so the number of separate activities will tend to multiply.

In many instances, the effort needed to coordinate, control and monitor these disparate activities may become a serious barrier to organizational efficiency and the development of innovation. Unless carefully managed, organizational complexities and politics will strangle innovation and creativity. Conversely, in situations where the organization structure is less rigid, the interfaces between these various activities will often become the single biggest source of wasted effort, misunderstanding and interdepartmental suspicion.

There is no wonderful solution to the problem, unfortunately. However, there are ways of proceeding which can help to reduce the difficulties, and much of this book is concerned with describing these. In this chapter, we first look at the role of engineers within the organization with regard to meeting objectives and achieving competitiveness. We then examine how

1

the engineer's performance may be undermined by bad management of interfaces between engineering and manufacturing, and engineering and marketing. Finally, we look at the tensions which can be further imposed by the financial structure of the organization, and the organization's culture.

Engineering and organizational objectives

All organizations are unique in that they have differing purposes, technologies and objectives, and will structure themselves and their activities in an individual fashion. Much can be done to reduce the organizational problems mentioned above, if the whole of the organization is working to achieve clearly identified objectives. For example, the DSS now publish targets for the maximum time it should take to process and pay benefit claims. Software engineers are required to develop software which is capable of dealing with the volume of claims, and the information required to assess each claim, within the allotted timescale. Administrators and DSS officers will similarly plan their own activities.

In a commercial organization, the prime objective is to achieve a satisfactory return on the capital invested in the business. This is an important measure of the vitality of a business, since it would be pointless to invest one's life savings in a risky project in the hope of gaining a small profit, when greater returns could be made by depositing the money in the National Savings Bank with no risk at all. The measure of business performance is referred to as 'Return on Capital Employed' (ROCE), and is developed in greater detail in Chapters 6 and 7. A firm can only achieve a ROCE greater than the industry's average in one of the two ways listed below, and both commonly require important inputs from engineers.

Cost leadership

An organization following the strategy of cost leadership aims to have costs lower than any competitor. This means that at a given market price, the lowest cost producer makes a greater amount of profit per unit sold than any competitor. Engineering activity will focus on designing products that can be made cheaply, and innovating within the manufacturing process to keep down production and distribution costs. The links between engineering and accounting would have to be strong in such a firm.

Differentiation

An organization following a differentiation strategy will aim to make its products significantly different to those of its competitors, by offering enhanced and extra features, after-sales service and customizations etc. Innovative engineering, coupled with the marketer's knowledge of what customers will be prepared to pay extra for, is clearly vital here. In such a firm, expenditure on research and development and marketing is likely to be high, so it is unlikely that the firm could also be cost leader as well as heavily differentiated, but high returns can still be made because customers will be prepared to pay more for the differentiated product. However, this does not mean that the firm can be sloppy over cost control, as there is little point in gaining extra revenues through differentiation if the money is going to be frittered away on frivolous projects or wasteful production methods.

Whichever route the organization takes, it is clear that the engineer's prime contribution is through ensuring the competitiveness of the

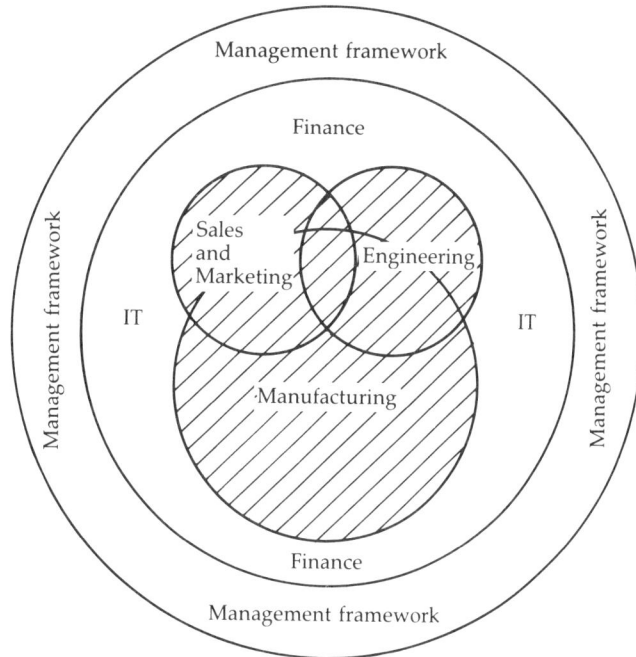

Management framework

Finance

Sales and Marketing

Engineering

Manufacturing

IT

IT

Management framework

Management framework

Finance

Management framework

Management are accountable to shareholders for a return on investment and providing leadership and reward structures to employees for optimum motivation

Communication methods include:

Financial and operational criteria

Figure 1.1 Engineering activity within an organizational context

Information technology to ensure fast effective monitoring of results by management and employees;
Sales and marketing, engineering and *manufacturing* produce and increase the company's wealth by providing customers with services and products for mutual profit.

company's products in the marketplace. This point is taken up again in Chapter 3. However, successful innovation in the engineering drawing office, or in production design does not, in itself, ensure a competitive advantage and above average returns on capital. In the remainder of this chapter, we examine the ways that interaction with other parts of the organization can undermine genuine innovation and competitiveness.

Engineering activity within an organizational context

Figure 1.1 illustrates the way in which engineering activity fits within the organization as a whole, and highlights some of the activities in which engineering makes a key, but not the sole, contribution. Engineers then, must first learn to carry out their roles within the department in which they are employed. But successful engineers must also learn to function effectively within the managerial constraints imposed by an organization's structure. This will include any strategic direction given to the engineering department following managerial perusal of financial and market indicators, and also the status and authority given to engineering departments within the corporate culture. The engineers may have more enhanced status and authority if the company believes itself driven by technological innovation,

3

than in an organization driven primarily by marketing and financial considerations. Moreover, the engineer must learn to manage the legitimate overlaps of activity between design engineering, manufacturing engineering, accounting and marketing.

The interface of design engineering and manufacturing engineering

Both manufacturing and design engineers have discrete responsibilities and duties. Qualified engineers with relevant specialisms in design, software, robotics and manufacturing systems will be found in both areas of activity. There are many important daily communications and tasks involving both design and manufacturing engineers, including, among others, engineering drawings and attendant documentation, design changes, research and development and prototype development. These important links between the two areas are discussed below.

Research and development and prototype development

Once the buyer's needs have been established by marketing, the research and development (R and D) engineers will then design and build a product which satisfies those wants in a way profitable to the organization. At an advanced stage in the development process, a prototype model is normally built in the factory (a process which can itself cause disruption on the shop floor if not properly planned).

Prototypes are assembled, as far as possible, from standardized components made in-house or already available from current or new suppliers. An experienced draughtsman will attempt to reduce the number and variety of components used to build the prototype, in order to control component costs and minimize manufacturing complexity from the outset. Since the organization's purchasing decisions will be based on these drawings, it is vitally important that the designer and draughtsman cooperate efficiently from the outset of the project.

In addition to specifying the components required for the prototype, it is often the case that research engineers' design will become a key influence on any capital investment decisions which have to be made in order that a successful prototype can go into full production (although most prototypes are not successful). The wrong choice of machines and other capital equipment is not usually reversible, and even if the organization can afford to replace a costly mistake, the delays involved may mean that a competitor has stolen a market lead, or a seasonal market has been missed. The marketing department should assist the engineer by providing sales forecasts, so that the engineer can estimate the time taken for the project to recover the necessary capital investment (which we refer to as the payback period). The engineer can then design the optimum combination of machines and manufacturing systems in the production process, on the basis of informed projections rather than rule of thumb or guesswork.

A further series of constraints may be imposed on the engineer's manufacturing design by quality control, who are responsible for best practice and the elimination of product defects on the production line. They may require specialist monitoring equipment for statistical process control (SPC; see page 98). The drive for quality management will extend to ensuring that the right skills are available on the shop floor; and that training and retraining of the workforce will remedy any skills deficiency before the prototype goes into full production. Early consultation between quality control and R and D is therefore crucial.

Drawings and documentation

Once prototypes have been passed, the product passes into full production. Of prime importance are the master drawings produced, maintained and modified by designers, in either a hard copy library or a database. Without these drawings, nothing can be built on the shop floor. Furthermore, these drawings are used to determine the supplies of components and other inputs that will be required in the manufacturing process. These requirements are communicated to manufacturing and the organization's purchasing departments via a document referred to as a 'Bill of Material' drawn up by the designer. The master drawings and the Bills of Material must be continually updated and maintained by the designer to avoid communicating mistakes or outdated instructions to other parts of the organization, which will act, committing resources, or make decisions about costs and prices which will affect the product's potential success in the marketplace.

Design changes

In many cases, designers will be working to a specification derived from customer requirements. Often, the initial design of a new or modified product may satisfy customer specifications, but the design may fail to acknowledge the realities of actually making the product on the shop floor.

A prime activity of production engineers is job planning, which includes providing feedback to the designer on design errors, as production engineers are not permitted to change the design themselves. Since job planners may be working to tight schedules, there can be a continual source of friction if there is not a fast and effective system of incorporating this valuable feedback into the design through a design change order (DCO).

Naturally, when designers have an awareness of the challenges and difficulties in job planning, job tooling, machinery and manufacturing systems (including flexible manufacturing systems, see Chapter 5), they are in a position to save substantial labour time and minimize the number of DCOs. But it is vitally important that designers and production engineers cooperate to satisfy the original customer requirements by making best use of the most modern systems and technology available to the organization. This cooperation is a crucial means of ensuring that the organization maintains cost competitiveness on existing products, forming a sound starting point for effective product development.

Product liability

Once the product is actually launched, further quality or performance deficiencies may come to light, usually through a customer service department. An engineer is often required to diagnose the problem, and take corrective action personally or, through sales or manufacturing personnel, is called out for field modifications. Hence, the concerted effort of manufacturing, service, quality and design departments may be necessary to win repeat orders. This activity may take up a substantial part of engineering time, and is generally non-productive time, since it is not adding to competitiveness or the long-term goals of the organization. Domestic and EC regulation makes product safety an ever more important issue.

The interface of engineering and sales and marketing

Sales and marketing are primarily concerned with aligning the organization's resources toward satisfying customer wants in the most profitable manner. The means by which this is accomplished are discussed in Chapter 2. Engineers are foremost among those whose activities must be aligned toward the organizational goals if the salesforce is to have worthwhile products to sell competitively at a profit. Communication will occur in many areas of activity, including daily commercial interfacing, forecasting, budget planning and in marketing individual products. These areas are discussed below.

Daily commercial communications

Engineers are required to assist in the preparation of quotations and tenders in response to customer enquiries. They are able to provide realistic cost estimates for bespoke products and product modifications, and furnish realistic dates for design completion. This day-to-day activity is vital for ensuring future orders, by monitoring changing customer requirements, upturns and downturns in the volume of business. But these activities can easily swamp a design office, drawing attention away from innovative projects necessary for the organization's long-term development.

There is an ongoing need to monitor the organization's effectiveness in converting expensive design effort into worthwhile orders, particularly in a business downturn, when management is under pressure to save effort on spurious enquiries. However, the day-to-day requests from sales and marketing will generally conflict to a greater or lesser degree with longer term strategic development work originating from the same source. Balancing these conflicting priorities is a major challenge to management.

Engineering resources and budget planning

Marketers pay particular attention to gross profit derived from individual products and product ranges. Profit trends are analysed annually and monthly, and resulting marketing decisions may have managerial implications for engineering.

The monthly marketing analysis is undertaken to detect significant changes in the sales of the entire product range (or product mix), in order to detect a fall in the sales volume of any product which is a major contributor to the firm's profits. If a downward trend is identified, reversal becomes an immediate priority and engineers may need to make necessary design and product modifications.

The annual analysis examines changes in sales and demand at the market level, with the intention of providing data for forecasting trends in the whole of the market, rather than the performance of their own products in each market. The purpose of the exercise is to establish objectives and priorities for the coming year, for example, a company may decide to curtail R and D on a successful product in a declining market, in order to focus activity on a rapidly growing market that the firm has not yet entered.

Following both monthly and annual analysis, development managers will have to make rapid adjustments to the deployment of their staff on existing projects and the allocation of departmental budgets. It may also become necessary to recruit or hire contract engineers with new or specialist skills.

*Marketing of
individual products*

Both marketers and engineers continually assess their products in terms of strengths and weaknesses, the opportunities to develop their products and increase sales, and possible threats to current and future value to the company (this process of examining strengths, weaknesses, opportunities and threats is normally referred to as 'SWOT' analysis and is discussed in more detail in Chapter 3). SWOT analysis may help marketers develop strategies for their products, which may involve establishing performance and quality criteria for a given price, copying and improving a competitor's product (referred to as 'Me too Plus'), or trying to identify a new niche in the market.

Once the strategy for the product has been defined, engineers will be heavily involved in all aspects of product development, and will be expected to contribute to promotional strategies by preparing graphics and product feature details to be included in promotional literature and service manuals. Engineers may also need to answer queries from distributors and end users directly, in a professional manner, otherwise loss of business is very likely.

While the marketer is preparing a pricing strategy, the engineer will be required to design to tight cost targets and provide crucial data on product performance, content and quality. This information will be considered together with similar data for competitive products. Marketers will set a price which is competitive in the marketplace, but takes into account any premium that can be made from a superior product.

Sometimes the interface between the two functions is eased by the presence of engineers on both sides. The title 'Sales Engineer' is now common, and there are a few reported sightings of 'Marketing Engineers'.

Case Study 1.1: Land-Rover

Mystique of market leader

How long can Land-Rover go on delivering the goods? The four-wheel-drive favourite has defied all predictions by remaining consistently profitable throughout the recession. The range consists of the perennial Land-Rover itself – now renamed Defender – the luxury Range-Rover, and the Discovery, a runaway success since it was introduced in 1989.

It has seen off a succession of rivals, including mighty Mercedes. Each time a competitor appears, it just seems to expand the market, in the process identifying fresh potential customers for Land-Rover. But although its present line-up is selling well, its privileged position cannot last for ever. Land-Rover's biggest asset is its brand name and experience. But it is tiny – Solihull turned out just 53,000 of all three models last year – and two of its models are too old to benefit fully from the new techniques that have transformed modern car manufacturing.

One of the Land-Rover models was designed for easy manufacture, and key to product quality and factory productivity. Amazingly, the

Defender has had just one redesign since its launch in 1948, and even the Range-Rover dates back to 1970.

Until recently, the range was starved of the investment needed to turn the modest success into a big one. As the Land-Rover niche attracts more attention from the large manufacturers for its recession-beating qualities, the company badly needs investment in a new product to replace its older models. It hasn't happened yet, but one day a rival will combine manufacturing resources and a winning design to produce a world beater. And at that point Land-Rover will need more than its legend, marketable name and performance track record to continue its run of success.

Sunday Express, 24 January 1993

The idea of project management

There is an absolute need to manage these interfaces effectively, since the activities involved are of fundamental importance to the success of any project which requires change. However, traditional departmental structures are not best suited to this managing of change, organized, as they are, to effect routine procedures using an established range of specialist skills and interests. Manufacturer are increasingly relying on multi-disciplinary task forces, with appropriate skills selected company-wide, to manage individual projects across traditional departmental frontiers. Consequently, many engineers now operate as project managers, striving to achieve a balance of sound engineering production practice, R and D flair and business acumen.

The engineering department within the organization

In addition to the interfaces between engineering and other functions described above, engineers also work within financial constraints and the corporate culture as a whole. Below, we consider both in turn.

Engineering and financial constraints

Engineering and its related activities are ultimately resource users (in particular, capital, manpower, materials and machines). Resource usage is generally measured in financial terms. In this respect, all business activities will be subject to the financial disciplines imposed by the organization, as determined by the accountants. Accountants and engineers share similarities in that they are both widely recognized as distinctive, functional and fully fledged professionals. Their status is due in no small part to the specialist academic education and training each receives when acquiring formal professional qualifications. During the lengthy qualification period, both may fail to recognize the importance of professional skills outside their chosen vocation, or appreciate the impact the disciplines will have on each other when carrying out their respective roles in the future.

Yet this interdependence becomes self evident when we consider the increasing complexity of many organizations, and the dynamism of their operating environments. Senior management may attempt to cope with this

Figure 1.2 The place of project management in coordinating multi-disciplinary enterprise

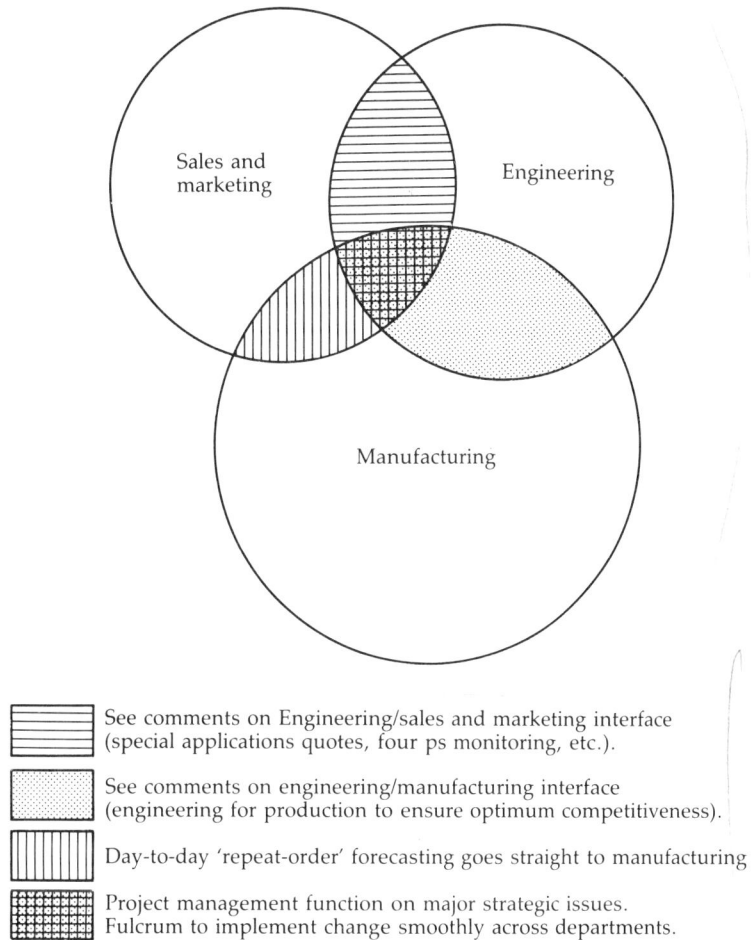

See comments on Engineering/sales and marketing interface (special applications quotes, four ps monitoring, etc.).

See comments on engineering/manufacturing interface (engineering for production to ensure optimum competitiveness).

Day-to-day 'repeat-order' forecasting goes straight to manufacturing

Project management function on major strategic issues. Fulcrum to implement change smoothly across departments.

complexity by setting measurable managerial objectives to plan and control business activity. These objectives are commonly set in financial terms (or have a substantial financial content at least), and will include profit, revenue and cost targets. For example, a construction engineering firm may be expected to achieve a certain profit target, or rate of return on its capital, to demonstrate competitiveness and attract future investment.

Although this target will be set at the highest managerial levels, a cascade of sub-objectives will be required at every managerial level, including engineers as project managers leading multi-disciplinary teams. By these sub-objectives, the activities and priorities of the engineers will be determined by financial criteria, understanding of which requires study or experience outside their technical background. Similarly, an invitation to bid for a highly specialized machine-tool contract may seem to involve a straightforward technical assessment of the development and production costs involved in fulfilling the contract. A successful bid will guarantee work for the company concerned, but will it be a profitable activity? And if so, will the profit made be a better return on investment than other possible projects open to the company (or better than depositing the investment capital at the Post Office), and over what time period will the profit be made? Are the physical and human resources already available, or will the project require the company to borrow, lease, recruit, train or sub-contract? Will large sums of cash become tied up for a long period of research and development? Should the company be bidding at all? In short, the technical implications of the bid will have financial implications which have to be considered in the context of the firm's present financial position and planned financial objectives.

The example above illustrates the need for financial planning, control and decision making to coordinate activities in engineering, and related activities, to support the overall financial requirements of the firm. However, the subordination of engineering activity to accounting activity can be misconstrued and resented by engineers, particularly when the firm is operating at less than full capacity.

However, it is entirely misleading to consider the engineer as a somewhat helpless technician, completely bound by the operational procedures of business strategists and accountants. In any manufacturing organization, it is the interdependence of engineers, accounting and marketing which is of prime importance. Although engineering may be directed by marketing and accounting, the status of engineers should not be correspondingly diminished or the importance of their contribution overlooked.

The place of engineering in corporate culture

The status of the engineer within a particular organization will be influenced by the way in which the organization perceives itself. The corporate culture of many engineering firms places the engineer at the centre of all commercial activity and sees marketing and cost control as supportive functions. On the other hand, other firms regard themselves as driven by customer needs, and therefore emphasize the marketing and accounting functions. A full account of the ideas and implications of corporate culture is largely beyond the scope of an introductory textbook. The authors wish merely to contrast two model types, engineering led and marketing led.

Marketing led versus engineering led

First, an engineering, (or technology) led organization believes its strengths lie in the way it can quickly exploit its research and development activities by bringing innovative products and product features to market. Companies operating in such high technology markets as fibre optics, holography, bio-technology, computers and medical scanning machinery may well be invention and innovation driven. In such firms, the research activities of the engineers are prime, and the salesforce will generally be composed of engineers or those trained in some related technical discipline.

The second model is that of the company directed by marketing activity, that is to say, a company which builds its strength on identifying and predicting changes in customer requirements quickly and accurately, and incorporating these developing customer specifications into new and existing products. Such companies will aim to determine the dynamics of their market, and will develop strategies to maximize sales of standardized products in the most profitable markets. The factors which contribute to the development of these strategies are discussed in Chapter 3; but clearly the scope for engineering activity will be curtailed by financial

---

Advantages of marketing-led companies

- Can prioritize product development towards more profitable markets.

- Can plan long-term investment programmes which take into account shifts in consumer preferences.

- Can control profit margins on each job through greater understanding of what the buyer actually wants, how much the buyer is prepared to pay and what the competition is offering.

- Can focus on full back up to ensure repeat orders.

Advantages of technology-led companies

- Product uniqueness, particularly in terms of new products and new applications.

- Ability to specialize for small niches.

- Development budgets likely to be given priority.

Disadvantages of marketing-led companies

- Problems in low volume and bespoke marketing.

- Tight cost control may limit product ranges.

- Well intentioned marketing research may stifle original thinking by innovators.

- Tinkering with marketing applications of existing technology may enable competitors to steal a lead

Disadvantages of technology-led companies

- Poor marketing intelligence may lead to underpricing and poor marketing generally

- Disjointed product ranges; including many products which no longer make a profit

- Reactive approach to market trends

- Low priority given to cost estimation, making pricing and capital investment decisions somewhat imprecise.

Figure 1.3 Marketing-led and technology-led firms contrasted

The pros and cons of the two approaches contrasted

considerations, and the need for a sales engineer is largely confined to the in-field testing of prototypes and investigating customer complaints. The two models are contrasted further in Figure 1.3.

In reality, of course, most firms exhibit some characteristic of both types, although the current prevailing wisdom suggests that firms driven by customer wishes are more likely to prosper than those led by their products. However, there are clear advantages and problems associated with both models. It is likely that engineering-led companies would work better in a new, technological market, where there is little or no competition. The company would then be in a better position to dictate selling price and delivery terms for as long as its technological lead could be maintained. On the other hand, the engineering-led company may not be able to compete successfully in a mature market, against competitors who accurately predict consumer wants and provide a range of products, each suitable for a potentially profitable sector of the market.

Case study 1.2: Company K

A company in flux: changing from technology leadership to marketing leadership

Company K entered the domestic showers market spectacularly in 19X8, when their founder patented and produced an affordable and safe solution to a long-standing problem; that of using electricity to power a shower without any risk of electrocution for the user. The founder and his associates risked their savings and property in setting up a production line, and the product was immediately recognized by consumers as a major improvement on existing shower systems. Large retailers, including the Electricity Board showrooms, adopted the product and used their ready-made distribution system.

In spite of the initial successes in production, the firm quickly experienced difficulties. In particular, the retailers dictated both the market price to consumers and the profit margin enjoyed by Company K; they then complemented the narrow product range with items made by Company K's competitors in both home and overseas markets. Moreover, in spite of profitable trading and rapid expansion, the company began to experience difficulty in paying its suppliers (an often fatal process referred to as 'overtrading').

The firm appointed a well known business strategist to join their board of directors as chief executive. He immediately set about putting the company on a firmer financial footing by going public (see Chapter 6). The distribution operation was reorganized, so that Company K took the lion's share of the profit, rather than the distributors. The founder and his original associates continued to organize and develop the technical innovations which led to the firm's initial success, but had accepted the need for strategic marketing and business planning at the centre of their activities.

Chapter summary

An engineer has a discrete range of technical skills that can be applied to a range of business problems. Prime among these is the importance of maintaining competitiveness in a commercial organization. But increasingly, the engineers must interface with a wide variety of non-technical specialists within a business organization. Moreover, engineers, like other specialists, must work within financial constraints and the prevailing corporate culture. Managing these interfaces is intrinsically difficult in a traditional departmental organization, regardless of whether a firm believes itself driven by the demands of its customers or the quality of its engineering innovation. Multi-disciplinary project teams would appear to be an effective alternative which could go some way to combining merits of marketing intelligence and engineering innovation.

Tutorial exercises

1.1 To what extent will the future of Land-Rover be determined by its past and present engineering activity?

1.2 How well did Company K exploit the advantages of being engineering-led?

1.3 To what extent were the possible pitfalls avoided or minimized?

Case study 1.3: Peter Smith

The story of a sole trader

Peter Smith had been an engineer for more than thirty years when he accepted a generous redundancy offer. But he was disappointed when he found difficulty in getting another job, and began to consider alternatives.

He eventually used his knowledge to design a range of specialist parts for the automotive industry. When he showed the designs to his many acquaintances among the small engineering firms in the locality, he was encouraged to set up his own firm to produce high quality, customized components.

His personal savings and redundancy money were not enough to get things moving, so he borrowed a substantial sum from his bank to buy machinery and other capital items. His bank manager was very enthusiastic, and helped him to fix up a deal with the Westing Development Corporation, whereby he gained a workshop rent-free.

The first few months were hectic. He found he had to make the products, load and unload lorries, keep records and accounts, look for new business and dozens of other things that he had not anticipated. He took on eight workers, including his wife, who gave up her own job and took over all the administration. Eventually, his business took off. The products were everything his customers had hoped for. Repeat orders and new customization orders for around £2000 to £8000 each, gave Smith more work than he could handle. He employed extra skilled workers on the production side, but his administration had difficulty keeping up.

Just as he began to get on top of the business, the rising acrimony between unions, management and government over the future of the steel industry developed into a major national strike. Smith had difficulty getting supplies at first, and, as the strike continued, he found that his customers began cancelling orders. His suppliers began pressing him to settle his bills, but his own customers were not paying up. Several defaulted on payments. His cash flow dried up, and his debts began to look unmanageable.

In a last attempt to save his business, he approached the larger national firms in the industry, as these had not been so badly hurt by the strike. He immediately won a major order for £200,000 – which would have solved all his problems – but his premises and production systems were too small. He approached his bank for a further loan, and they promised to think about it. But an unpaid supplier sent in bailiffs to take away his machines and stock before the bank had made a decision.

Smith was declared bankrupt. In addition to his business debts, his overdraft was £16,000 and his wife had debts of £4,000. The Westing Development Corporation pointed to the small print in his contract for the workshop and demanded £7,000 back rent. All Smith's possessions were sold off – except his clothes, bedding and engineering tools.

1.4 (a) With reference to Case Study 1.3, what were Smith's objectives in setting up his manufacturing business? Were there any alternative ways of using his skills and experience to achieve the same ends?

(b) What market research did Smith carry out before setting up his workshop? What was overlooked; with what consequences?

(c) What alternative sources of finance could Smith have considered, both in setting up the business and managing his cash flow?

1.5 BMW recently claimed to have developed a prototype motor-car engine capable of running on both petrol and liquid hydrogen. The liquid hydrogen will be produced by a solar-powered process. Why is there a need for such a car? What engineering and marketing challenges will be faced if the prototype is put into full production and sold to consumers?

2 An introduction to marketing

In Chapter 1 we briefly alluded to the importance of ensuring that engineering activity was focused on producing products and services which satisfy customer wants profitably, rather than directed at product improvement for its own sake. This orientation towards satisfying customer wants is commonly referred to as the marketing concept.

The marketing concept

The Chartered Institute of Marketing defines marketing activity in the following way:

> Marketing is the managerial process which identifies, anticipates and supplies consumer requirements efficiently and profitably.

Although this is by no means the only way of defining marketing, this definition is more than a useful point of departure. In particular, it is worth noting that:

- It is not something that professional marketers do, it is central to the activities of engineers, accountants and managers of all disciplines.
- That marketing activity is not constrained by the need to promote an existing range of products to an existing set of customers, since it anticipates as well as identifies customer wants.
- That marketing should be carried out profitably, which implies that the role of the engineer in creating competitiveness by cost reductions or differentiation is prime.

In this chapter, we develop the first observation by looking at the way in which marketing orientation can be expressed, both by marketers and engineers. In particular, we examine the four major tactics available to a marketer; the product itself, and the pricing, promotion and distribution (also known as 'place') of the product. Collectively these are known as the marketing mix (or four Ps).

In Chapter 3 we examine the second observation, and its implications for planning marketing activity. In Chapters 8 and 10 the engineering implications of the third observation are examined in some detail.

Thinking marketing

In many respects, it is not immediately obvious why engineers should continually think about customers and markets, since their day-to-day activity and training tend to require concentration on immediate, practical and concrete technical problems. But clearly, the whole of commercial activity is based on the notion that there is a consumer who has a need, and will place a value on any product or service which satisfies that need. It is helpful then, to think of a business and its associated activity in terms of the value it creates for its consumers, rather than the products and services it sells. The process of creating value is said to have three major stages, outlined below.

The stages of value creation

Stage One

Identify the value to be created, by analysis of customers, their needs and what they are already offered in the marketplace.

Stage Two

Make the value by designing products which offer the value identified in Stage One, develop appropriate after-sales and back-up services, make, price and distribute those products.

Stage Three

Express the value created in Stage Two by advertising and promoting the products and services.

Value creation is not sufficient to keep a firm in business. If other firms are creating the same values more cheaply, or offering better value for the same price, an organization will be uncompetitive. In Chapter 1, it was pointed out that the firm can only become competitive, and remain so, in one of two ways. A firm can produce its value in a way which its consumers will see as distinctive, either in terms of the product offered, customer services, or availability. Many successful firms gain a competitive edge in this way, by product differentiation. Alternatively, a firm can become a cost leader.

Engineering skills have a vital role to play in the process of creating value, and it is worthwhile examining the way the process appears from

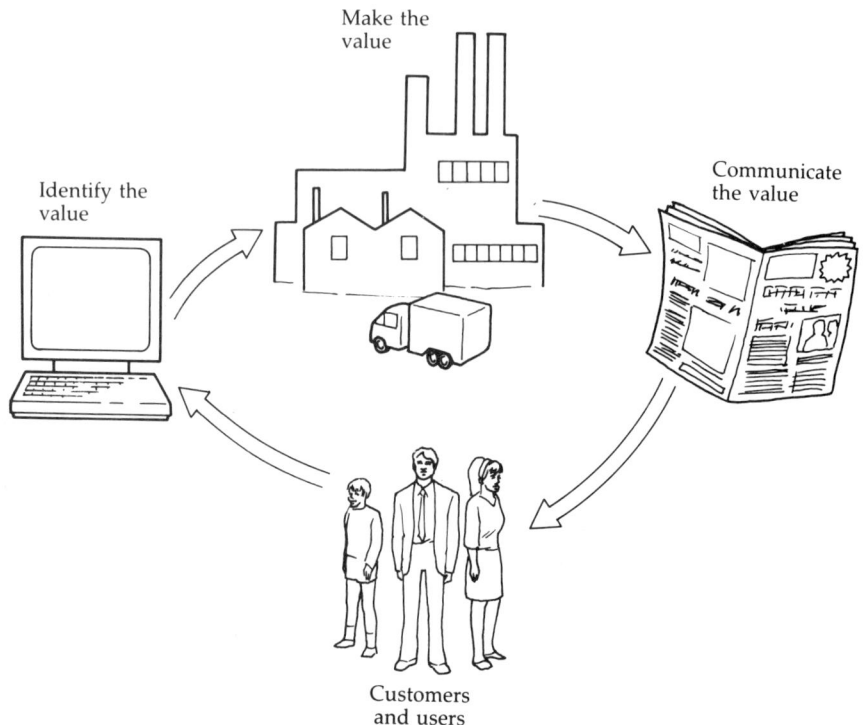

Figure 2.1 Stages in the creation of value

the viewpoint of a PC or bench-bound engineer. Clearly, a sedentary engineer will focus entirely on the second stage, where technical skills are concentrated, and rely on sales and marketing professionals to sell the product created. Products will only change and develop according to changes in technology available to the engineer. In a difficult situation, the engineer will be required to focus on cost cutting problems, and the firm will increasingly rely on harder selling and promotional activity (i.e. Stage Three activities).

This emphasis on Stages Two and Three of the process is sometimes referred to as the selling concept or sales orientation, since management is generally orientated towards selling more and more of their existing products. In a firm operating under this system, those engineering and selling skills identified as part of Stages Two and Three are absolutely vital to the success of a firm.

Although sales oriented firms can be very successful, there is a heavy reliance on good luck and managerial instinct to align the skills of the company towards producing and selling products that consumers actually want to buy. But by disregarding the activities in Stage One, a firm will only develop products which consumers wish to pay for by chance, and then only until a product appears which is more carefully targeted at the customers' needs. There are many examples where dependence on engineering innovation and managerial intuition has led to failure, including the first digital wrist-watch, which required both hands to operate, the first commercially produced electronic motor-car, which was unable to perform to acceptable standards, and the first aerobatic kite, which was relatively complicated for the consumer to assemble, and only available in plain blue. Success is by no means guaranteed by examining customer wants and attempting to create appropriate value, but consistent, long-term success requires that consumer satisfaction should be attained systematically rather than as an occasional or fortuitous occurrence. Clearly then, those activities identified in Stage One must set the parameters in which the engineering and promotional skills of Stages Two and Three are applied.

Stage One: Identifying needs and the opportunity to create value

Since every individual person and every firm are unique in many respects, it follows that the needs of each potential customer are similarly unique. Consider the case of software designed to replace a bureaucratic paper-based process. A large organization may decide simply to reproduce the existing system electronically, or may take the opportunity to redesign the system completely. In either case, there are major opportunities for software and systems engineers to direct their skills to the particular needs and problems of that client, and the finished product would be significantly different. Alternatively, the company might be persuaded to buy an existing system, with sufficient flexibility to accommodate the idiosyncrasies of the buyer. This would be a low cost option, and the only option available to a small firm.

Both large and small firms are looking for software systems, but it is not really sensible to think of them as being buyers in the same market. A seller will need to offer different services and products to these clients, and since the clients may well make different uses of it once it is installed, the after-sales service will also be different. Although the engineering skills used in the development of the products the two firms elect to buy may be similar, they will need to be applied toward achieving different ends for each buyer. In marketing terms, the software market is segmented into bespoke products (for large users) and 'off the shelf' products (for light, or unsophisticated users).

Market segmentation

Identifying and separating types of buyer in this way are known as market segmentation. The task of the marketer is to identify groups (or segments) of customers, and potential customers, with some common needs and characteristics. Once the segment, and its needs, have been identified, the marketer will work with engineers and others to derive ways by which these needs can be satisfied profitably by using the organization's skills and resources.

Markets can be segmented in a variety of ways, but the idea is most easily grasped when consumer markets are considered. A motor-car for example, satisfies a basic need of transportation. However, it is evident that a car suitable for a young, single person may not accommodate a large family; therefore there are two distinct market segments. Moreover, a young single person who uses a car continually, and has a need to impress clients, may be willing to spend a great deal for quality, status and comfort, while others may be less frequent users, have less money or simply choose to satisfy other needs. Clearly, there are many other segments within the original 'young single person' segment. Indeed, this must be so, because in reality each individual is in a segment of one, and the process of market segmentation is actually the process of aggregating individuals into groups with similar needs and characteristics. Given then, that there is no theoretical limit on the numbers of segments within a market, it is important to direct activity only to segments of sufficient size to generate a satisfactory profit.

Although each market will be unique, marketers often start by segmenting markets using the following criteria:

Physical location: locality, county, country, continent etc.
Usage rates: heavy, light, frequent purchase of small quantities etc.
Characteristics of customer:
Consumer markets: Age, sex, lifestyle, education, social class, income, expectations, personality, stereotype etc.
Industrial markets: market of final product, quality requirements and specification, service requirements, price sensitivity etc.

The customer characteristics section is separated into consumer and industrial markets, since characteristics differ greatly between the sectors. It is generally agreed that industrial buying is a rational process, involving clear quality and service specification and negotiation between equally well informed traders. Consumers, on the other hand, will approach many of their purchasers with less consideration of the alternatives, and less well developed expectations of the products concerned, than their industrial counterpart.

In many respects, the differences make identifying opportunities to create value in industrial marketing less difficult than consumer marketing, insofar as industrial buyers are able to articulate what they want, and assess the price they wish to pay for it. A marketer should be able to rely on feedback from the salesforce, and conduct market research to build up an accurate picture of the market's needs.

Consumer markets are more problematical, particularly since most producers never actually meet the end user of the product. Information may be forthcoming from distributors, and disgruntled consumers may bring their grievances to the producer, but otherwise the firm must rely heavily on market research to identify customer characteristics. It is from such research that the much maligned psychographic stereotypes have emerged; such as dinky (double income, no kids yet), yuppy (young urban professional), yummy (young, upwardly mobile) and lombard (loads of money but a real dunce head) have emerged. Working on stereotypes derived from the data by marketers, sociologists and psychologists, it is possible to direct marketing activity to substantial lifestyle segments of any consumer market.

Competition and market segmentation

Having identified the value, it is then important to examine how the value is created by existing producers, and whether or not it is possible to create a competitive advantage. An organization which believes that it has identified the opportunity to create value, will need to undertake research of the market, the competitors and the resources and capabilities of the company. The following chapter examines the process at some length.

Clearly, it is important that the firm does not commit resources to a project where it cannot sustain a competitive advantage for long enough to recover costs and make an acceptable return on capital employed. The two principal means of generating competitiveness, differentiation and cost leadership, have already been outlined. However the idea of market segmentation allows a third possibility, referred to as a focus strategy. With this strategy, one particular market segment is singled out, and the organization specializes in meeting the needs of that segment. This is the strategy adopted by Cray, who, although in the computers business, do not attempt to compete with IBM or other producers of standardized products. They have focused on producing supercomputers, with a range of applications far in excess of those required by most commercial or domestic users, but indispensable in many scientific and computational fields. However, it is important to realize that focus strategies still depend on creating differentiation or cost leadership within their chosen market niche in order to sustain competitiveness.

Segmentation is clearly a process central to successful marketing, but is not an end in its own right. It is a process designed to help marketers identify who their customers are, what needs their customers have, the means by which these needs might be satisfied and whether or not it is financially worthwhile to satisfy them. The answer to the last question will largely depend on the capabilities and characteristics of the firm in question. Should the answer be positive, the firm must then direct its attention to the creation of the value which will satisfy those needs in a way which will sustain competitiveness.

Stage Two: Creating the value

The engineer has a key role in the creation of value, once the marketers have identified the specifications of a product or service necessary to satisfy customer wants profitably. The engineer's greatest contribution comes in interpreting sometimes imprecise specifications into a feasible hard design criteria, designing and manufacturing the product. These elements are considered in Chapters 4 and 5. In this section we look at the factors most important to the marketer, the product itself, distribution and the price.

The product

Given a product specification, derived in the main from an idea of what consumers are likely to want, the engineer is expected to apply skill and knowledge to the development, testing and production of a product which satisfies those requirements. However, the marketer and the customer will attribute greater meaning to the term than the physical make-up of the product, as illustrated in Figure 2.2.

A product can be imagined to have three elements, a core element, tangible element and an augmented element.

> *Core elements*: Essential attributes, i.e. those things which satisfy the basic requirements.
> *Tangible elements*: Attributes such as special features, quality, design and packaging which add value and distinctiveness to a particular brand or product.

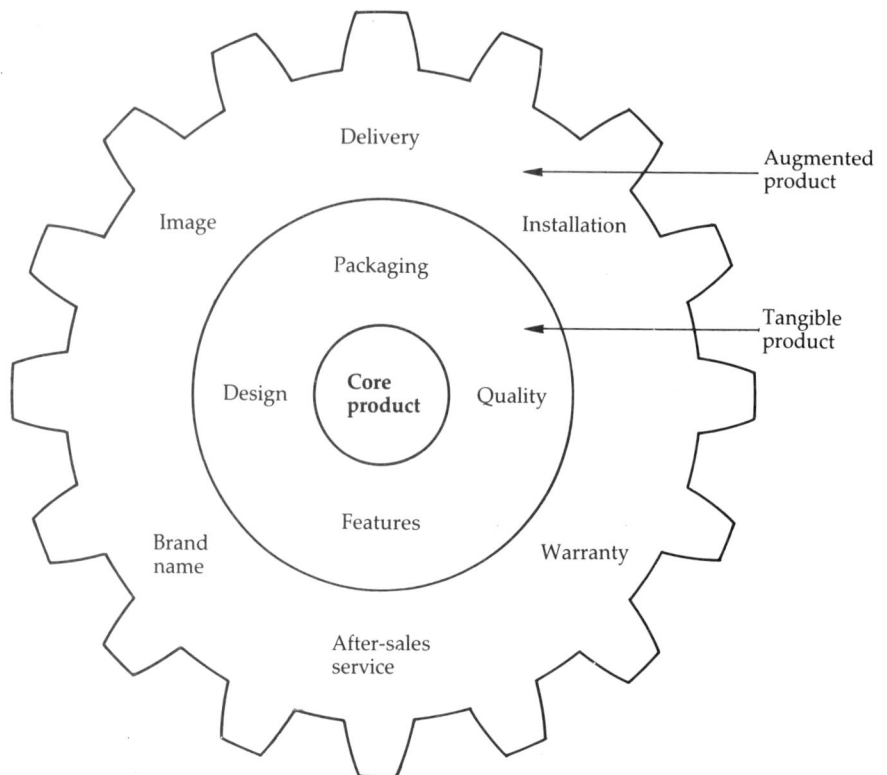

Figure 2.2 The product offering
Source: Palmer and Worthington

Augmented attributes: Attributes such as image associated with the brand, after-sales service, guarantees and warranty, delivery date and service, training and user support.

Product differentiation

The aim is to devise a product offering which is in some way distinctive in the marketplace. This aim is referred to as product differentiation, and is desirable because it creates value which is unique to that product or company. In this way, buyers may remain loyal to the brand in spite of lower prices or improved products elsewhere, in the short-term at least.

Engineers will be involved in all three elements, designing and producing the basic product, building in extra features and offering repair, maintenance and user support. Achievement of excellence in engineering will not, however, offset poor performance in delivery dates, warranties and after-sales service. And at the risk of repeating the obvious, such engineering inputs as extra features must be of value to the consumer to be worth building in, otherwise, as Apricot discovered with their PCs, the cost of designing and producing such features cannot be recovered by charging a higher price.

Over time, these extra features and design points may well become incorporated into the core of all competing brands and competition between brands will increasingly move away from these engineering inputs towards price, availability and image, where the engineer may have a less central role.

Distributing the value

In marketing terms, distribution covers the process by which goods and services move from the producer to the final user. There are two major aspects to the distribution problem, physical (or logistical) distribution and the channel design.

Physical distribution: Involving the scheduling, transportation and storage of the product. The engineer's role here is somewhat limited, although normally crucial. For example, the packaging problems associated with cigarettes in the UK are mostly promotional, in that the package must be distinctive and attractive. However, in distributing the same product in Chile, the engineer also must take account of the fact that the product will be handled less gently, and the journey may involve haulage by mule over the Andes.

Channel design: In industrial markets, it is not uncommon for the producer and buyer to do business directly. However, in consumer markets and industrial markets, a large number of intermediaries, such as wholesalers, retailers, agents and merchants will become involved. Again, the engineer's role will be somewhat muted, although in many industries, the saleforce will actually be composed of engineers, and as such, vital (hence the term sales engineers).

The purpose of intermediaries

Intermediaries add value to a product or service in a wide variety of ways, and release the producer to devote time to those activities in which there is distinctive skill or competence. For example, in the Peter Smith case study in Chapter 1, the proprietor was clearly able to create value in the

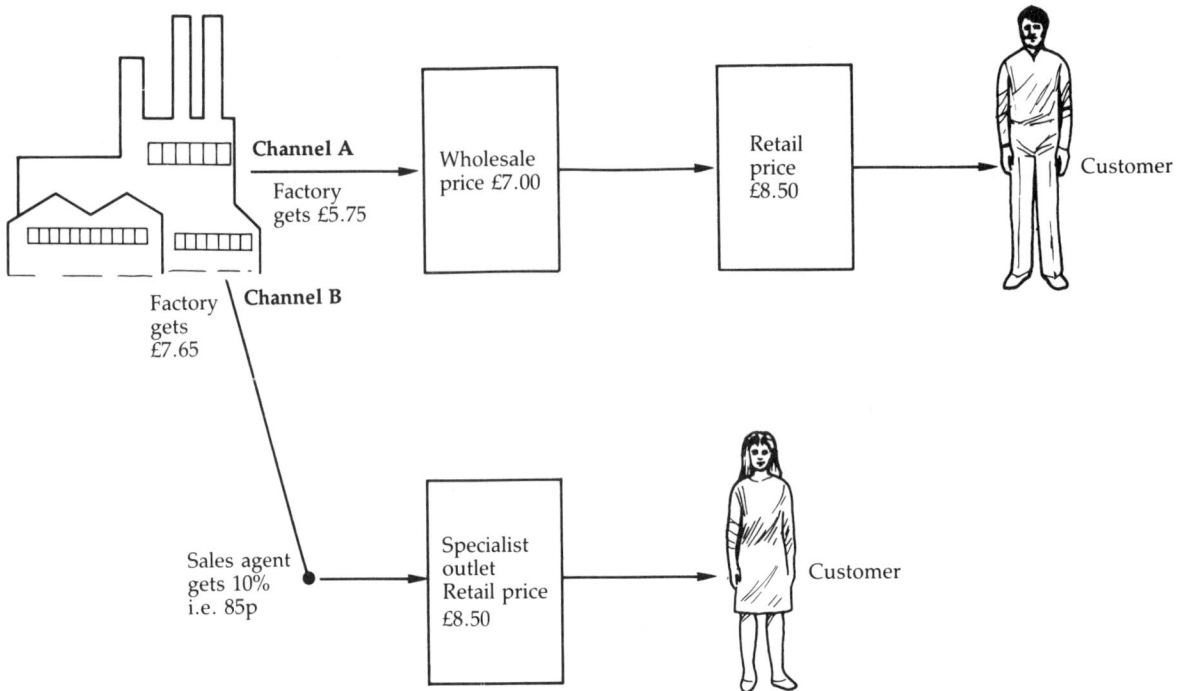

Figure 2.3 Distribution channels

design and manufacture of specialist, high quality products. However, his decision to do his own selling may have been unwise, since it meant that he had to spend much of his time 'cold calling' manufacturing firms who did not know him, and had no knowledge of his specialist skills. Perhaps a manufacturer's representative, or some other selling agency, could have added his designs to their portfolio of wares, exposing them to a much wider market than Smith could have achieved on his own. Moreover, Smith's time in selling would have been spent following hot leads only.

In addition to selling their skills and contact with a target market, intermediaries may also add value in other ways, including physical distribution, packaging and branding, after-sales service, provision of credit to the buyer, further manufacturing or processing, and, of great importance, providing market information to the manufacturer. In consumer markets, the retailer, for example, will add value by providing an appealing environment in which a shopper can be encouraged or persuaded to make a purchase. Indeed, some shoppers will perceive prestige department stores, such as Harrods, to add value in their own right.

Naturally, these intermediaries will require payment for their services, and the greater the quantity of services offered, the more will be expected. This is commonly achieved in one of two ways, first by means of discount and second by means of commission. In the first case, the intermediary actually buys the goods (or legally takes title of the goods), at a price lower than a notional retail price. In the second case, the intermediary never owns the goods, but takes a commission on the sale price actually agreed. The way in which this works is illustrated in Figure 2.3.

Channel B generates greater revenue for each sale made, since the distribution chain is shorter. But this is not equivalent to greater profit, as Channel B will incur a larger number of costs than Channel A, including maintaining a greater salesforce, paying for physical distribution and administration of a larger sales ledger. The additional expenses in Channel B could only be justified in circumstances where the product was of a specialist, customized or personalized nature, such as a design consultancy, or where individual products or contracts were of sufficient value to justify individual attention. Markets would include construction contracts or shipbuilding, or, perhaps more dramatically, where profits in a particular market have fallen to the extent that distribution and producers cannot both make a worthwhile profit.

Clearly, where there are large margins involved, Channel A can use more distributors to saturate every marketing segment. But if margins are small, or the product is perishable, the distribution chain must be kept as short as possible.

Distribution and competitive strategy

For many companies, the main objective of their distribution system is to place their product on a shelf next to their competitors' products, so that competition on price and quality can take place. But innovative companies see distribution in a much less passive light, seeking out opportunities to distribute products. The success of TieRack, in the UK at least, owed much to intelligent siting of outlets. Similarly, the success of keg beer in the 1960s owed much to the fact that the producers of the product were able to take over many public houses, thereby restricting the distribution of more traditional alternatives. Other companies have succeeded by targeting the needs of the distributor, rather than that of the consumer. For example, heavy users of cola and lemonade include those who mix them with spirits and consume the mixture in a bar, causing problems of storage and disposal of the mixer bottle for the proprietor. However cola and lemonade bottles in bars have now been largely replaced by syrup-based substitutes, bearing some resemblance to the original, mixed with a fizzy base at point of sale, because this is much more convenient for the proprietor.

In this last example, both the product, and the way in which it is dispensed, required engineering skills at various stages when first introduced. Calls on innovative engineering are not uncommon when a market is first segmented in a new way.

Co-ordinating distribution activity

Since much of the distribution and selling activity will be carried out by intermediaries, the producer must go to some considerable lengths to ensure that distributors will stock and sell the product. Briefly, this can be done in three ways – pushing, pulling and co-operating.

Pushing: This strategy requires offering a high discount on price and heavy selling, to the distributor. The distributor then has the incentive to push the product onto other distributors, retailers and the end user. For small firms, or unknown firms with innovative ideas, this may be the only option, but it can lead to a situation where the distributor makes more money than the producer (see Company K case study in Chapter 1).

Pulling: End users are targeted by promotional campaigns, thus creating a demand for the product. Distributors will then seek to satisfy this demand by stocking the product, thus pulling it through the distribution chain. By establishing demand in this way, the discount offered to distributors might be lower, and once the need for high promotional expenditure has passed, a higher profit margin can be retained by the producer.

Co-operation: It is argued that best results can be achieved when distributor and producer act together to attain agreed sales volume and profit margin targets, rather than approach the situation as a conflict over discounts and the division of tasks. No doubt there is much to commend this practice, which is common in Japan, but requires the development of long-term, non-exploitive relationships which are less common in the UK.

These relationships are not alternatives, and elements of the first two at least are common in the distribution of most goods and services.

Power in the distribution channel

In practice, much depends on the relative strength of those in the chain. It is increasingly the case that large distributors and retailers take a stronger position than the producer. Unless a product is unique, protected by patent or some other barrier to imitation, and in great demand, the distributor is in a prime position to pick and choose among competing producers. Once again, the engineer's contribution to building competitiveness should not be overlooked. However, it could be argued that, given the relative weakness of the producer in a market, ensuring effective distribution is the single most important marketing task.

Pricing the value

Price is somewhat unique among the marketing variables in that it directly affects both revenue and profit. Although economists have devoted much thought to the development of pricing theories, in reality these theories are of extremely limited help in price setting. It is worth remembering that the price must cover the whole of the product offering, not just the core product, and different segments will be prepared to pay different prices for different variants of the product offering. Moreover, segments which can only be reached by long or complicated distribution chains, will also have to pay more for the same product.

Following our discussion of distribution, it is possible to argue that in most cases the price to the consumer will actually be set by a distributor, rather than the producer. Indeed it is desirable that this is so in some respects, since retailers, for example, are likely to have a better idea of what buyers are prepared to pay. But it would be unwise for a producer to completely relinquish control of price for several reasons:

- Downward pressure on price will eventually work its way back to the producer.
- Upward pressure may lead to a greater profit for the distributor, but this may not be passed on to the producer.
- High price and low sales volumes may prevent a producer from expanding.

If producers do take steps to manage their prices in some way, they must have at least a notional idea of what price would be acceptable. This acceptable price will be composed of the following:

The price customers are prepared to pay: This will depend on both the value buyers place on the product, and the prices of alternative means of satisfying the same want that are available. Thus the price of a competitor's products and substitute products must be taken into account. Clearly, a company which succeeds in creating a perception of greater quality or performance in a brand, will be able to charge a higher price, or alternatively, destroy inferior quality competition by closely matching the price. It is amusing to note that some of the counterfeit Levi jeans produced in workshops in the Far East are allegedly of better quality than the originals, but still use fraudulent labels to benefit from the Levi reputation, and are priced slightly cheaper than the originals.

The price the buyer is prepared to pay: In many markets, the consumer and the buyer are one and the same. However, in industrial markets, the purchasing decision will be made by a professional buyer. A buyer will pay great attention to the price of purchases which have a major impact on purchasing costs, but may well be less diligent in other areas. We have already looked at the relative power of intermediaries in the distribution chain, but this power may well be reflected in negotiations over price. For example, in Chapter 8, examples are given for the cost structures of three competing firms bidding for an electric motors contract. This cost information was demanded by the purchaser, who had set a limit on the profit margin it was prepared to allow its supplier. Similarly, government contracts commonly require such information to be made available, and thereby restrict pricing opportunities.

In many consumer markets, the product will be purchased by one person for consumption by another, for example, male grooming products are commonly purchased by women. The buyer's attitude towards price must be expressly researched in such cases.

The cost of the product: Clearly a vital element, since the price must, in the long term, cover costs and add something to profit. The major difficulty is that engineers, accountants and marketers estimate costs in a different way. This difficulty is examined more closely in Chapter 10.

Strategic implications: Pricing policy will depend on whether the firm is seeking to establish a major presence in the marketplace, or to make a quick profit before moving on to another product or market. These strategic planning considerations are examined in Chapter 3.

Case study 2.1: Borland

Obstacles beset PC manufacturer
Borland faces growing competition and a price war, writes Louise Kehoe.

Just a year ago, Borland International was one of the fastest-growing companies in the personal computer software industry. The Scotts Valley, California, software developer had cemented its leadership in the database management program market with the acquisition of Ashton-Tate, and expectations were high for several new products.

Today, with some of these products still yet to be delivered, Borland is facing mounting competition and its growth has ground to a halt. The company's stock, which reached a peak of $86.75 in January 1992, has been trading this week at under $18 after disappointing third-quarter results.

On revenues down 9 per cent at $104.3m, or $2.34 a share, compared with net profits of $7.5m, or 28 cents, in the same period last year. The results included charges of $34.8m for 350 job cuts announced in December, consolidation of facilities, and inventory write-downs, Borland is now engaged in a potentially crippling price war with PC software market leader Microsoft and other competitors as it makes a long-delayed entry into the market for applications programs that run with the popular Microsoft Windows software. The delays have cost Borland dearly, forcing it to sit on the sidelines as sales of Windows applications programs soared to an estimated $2.9bn last year, an increase of 238 per cent over 1991, according to a market study published this week by Dataquest, the US market research group.

To make matters worse, when Borland finally launched a Windows version of Quattro Pro, its spreadsheet program, in September, the company blundered by packaging the new program with an existing DOS version of Quattro Pro. This software was supposed to appeal to a broad range of users, including those who use either Windows or DOS. But Borland has since acknowledged that instead it caused confusion. It withdrew the package in December and relaunched Quattro Pro for Windows as a separate product at a steeply discounted price. Third-quarter revenues were reduced by $10.7m, the company said, because retailers would receive rebates on their earlier purchases of the Quattro Pro product.

The price battle began in November, when Microsoft invaded Borland's most prized territory, the market for data-base management programs, with the introduction of Access, a PC database program for use with Windows. Offered at an introductory price of $99, about one-fifth of the price of Borland's established database products, Access has taken the market by storm and Microsoft expects to sell 750,000 copies by the end of this month. Firing back, Borland has announced a new 'introductory price' for Quattro Pro Windows of $99, down from $495. The company also said that when it finally begins shipping Paradox for Windows, the first of two new database management programs, it too will be heavily discounted to $149 for 90 days.

'Borland is celebrating its two-pronged entry in the Windows market with this promotion,' said Mr Philippe Kahn, chairman, president, and chief executive. Financial analysts, however, are already counting the cost and lowering estimates for Borland's fourth-quarter earnings.

In the longer term, it appears the 'introductory prices' being offered by Borland and Microsoft could establish new expectations among customers for downward price trends like those for PC hardware. Software developers may have a difficult time persuading buyers to pay higher prices, analysts warn.

While the price wars continue, Borland is also battling in the courts with Lotus Development. Last year, Lotus won a copyright infringement suit filed against Borland in July 1990, forcing Borland to cut certain features out of its Quattro Pro products. Last week, however, Lotus charged that the revised versions of Quattro Pro still infringe its copyrights. Borland is confident it will ultimately prevail in the dispute. However, if Lotus wins they may be forced to pay damages amounting to millions of dollars.

Financial Times

Stage Three: Communicating the value

Once Stages One and Two have been correctly planned, it is necessary to inform the potential buyers and distributors of the value which has been successfully created. The three main techniques for achieving this are personal selling with a salesforce, advertising and sales promotion, although such techniques as publicity, public relations, corporate image management, sponsorship and many others are of increasing importance. Collectively, these techniques are known as the promotions mix. In this chapter we will examine advertising and personal selling in a little more detail.

Advertising

Advertising, like all other marketing activity, must be directed at a particular target market, with a particular objective in mind. A good advertising campaign is planned around the five Ms: Message, Media, Motion, Measurement and Money, outlined below.

Message

The message must be designed to achieve some particular objective. Messages are often constructed around the need to communicate something about either the quality, features or uses of the product or brand; the quality of the company providing the product and the qualities of the consumers of those products.

Product messages: Messages frequently attempt to communicate some detail regarding the product itself. For example, Hellmans' award winning mayonnaise advertising campaign successfully communicated the idea that 'ordinary' people could use mayonnaise with a wide variety of meals, and was designed to change the idea that the product was one to be eaten by 'higher' social groups, and then only with salad. In industrial markets, advertisements are frequently more technical in nature, and are often used to support salesforce activity by helping sales

27

representatives gain access to important buyers, rather than actually promote a sale itself.

Company messages: Advertising campaigns may try to communicate something about the company itself. Campaigns from Dunhill, Zanussi, Ariston and BR have attempted to communicate something about the quality of the products and service they sell, without necessarily referring to specific goods. Such an approach is desirable when a company has a wide product range in many different markets.

The effects on the consumer are important. Evidence suggests that a well-known brand name gives a buyer a degree of confidence when making a purchase, and afterwards, helps to reinforce the choice made.

Some companies have attempted to use advertising to sell messages regarding the organization's environmental or ethical standing. In many cases, these campaigns have seriously backfired, as environmental groups have checked and criticized the claims made, and Friends of the Earth even announce an annual Green Con award for the most misleading environmental message.

Consumer messages: Other advertising messages attempt to communicate something about the nature of the consumer, by inferring that it is consumed by people with a certain status, image or lifestyle. It is, therefore, a major tool to be used in establishing a presence in a particular market segment.

This third category of advertising messages is often applied to consumer goods, in markets where consumers are normally unable to discriminate between brands in other ways. For example, consumers of draught ale and lager are commonly unable to identify their favourite brand in blind testing; even when a very strong preference has been expressed. Thus, making great claims about product quality would not seem a promising style of advertising message. Much of the preference felt by consumers apparently relates to the branding of the product, rather than product qualities at point of consumption. Advertising messages take advantage of this by encouraging the belief that consumers of certain beers may be more intelligent, adventurous, street wise or sexy than consumers of competing brands. In this way, a brand is positioned in the minds of the existing consumers, and repeat purchases and brand loyalty are established and maintained. Or, to be rather blunt, if advertisers fail to remind a consumer that he or she habitually buys a particular brand of beer or cigarettes, the consumer will tend to forget and switch brands. Evidence from tobacco markets, where advertising is restricted or banned altogether, tends to support this view (knowledge of this does not much help the consumer, or reader, avoid the effects of such positioning).

Objectives of advertising messages

The objectives of the message are commonly described in terms of the Attention, Interest, Desire and Action (AIDA) it can generate.

Attention: The message must gain the attention of the target market. This alone may be sufficient for a company wishing to retain its positioning in the mind of existing buyers. However, for more aggressive or informative campaigns, retention of the message is important.

Interest: Most companies will also wish to retain or instigate interest in the product, usually by relating product attributes to particular needs and wants, and by using humour, catch phrases, technical specifications or some other 'hook' appropriate to the target market.

Desire: Here, the intention is to create positive feelings about the product, company or consumer image advertised. It is, presumably, this feeling which is the contribution of advertising to maintaining brand loyalty and positioning; although in itself it is unlikely to offset bad feelings caused by poor quality, bad design or feeble after-sales service.

Action: Here, the intention is to get the target audience to do something. However, it is rare that an advertisement will actually instigate a new purchase (as opposed to repeat purchase). Advertising is helpful when encouraging potential buyers to seek further information, accept a trial offer or participate in some sales promotion, but, again, is unlikely to compensate for poor product performance (see Measurement below).

This AIDA model is widely used in setting advertising objectives, but is not without its critics. In many markets, such as industrial markets, the purchasing decision may be much more rational and less emotional than suggested by the latter stages of the model. Similarly, with Fast Moving Consumer Goods (FMCGs) like chocolate bars and soft drinks, many consumers think the product too trivial or frivolous to develop any deep emotional loyalty to it, particularly when the effect of a poor consumer choice is so small and short lived.

Media

Common advertising media include newspapers and magazines, trade and professional journals, billboards and posters, cinema, radio and television. When selecting media, the advertiser will first consider the target market. This requires not only some understanding of the reading and life-style habits of the target market, but also knowledge of market coverage, meaning that a number of local newspaper advertisements might be more appropriate than a national newspaper in such a market as car dealing, which has a strong regional bias. Second, the advertising message will preclude some media; detailed technical information is not well expressed by radio or television commercials, but is suited to magazines or special interest journals. In the case of a motor-car, Sunday newspaper magazines often carry technical details, supported by high impact television and poster campaigns. Third, the advertiser will consider the cost of an advertisement, normally in terms of the number of exposures to the target market per pound.

Motion

This concerns the scheduling and co-ordination of promotional campaigns, so that the advertisements and special offers run while supplies of special offer packs, free gifts or whatever are available. Further, the advertiser must decide whether an advertisement is to be shown repeatedly for a short time – to saturate the media – or less frequently to prolong the impact.

Measurement

How will the advertiser know whether or not the advert has been successful? It is rarely possible to gauge advertising effectiveness by reference to sales figures. Even if a somewhat gullible reader were to be motivated by a newspaper advertisement to visit a retailer, it may still be the case that the retailer is offering other products at a discount, does not sell the particular brand advertised or displays equally convincing point of sale advertisements for a competing brand. Thus, in this case, an effective advertisement would not lead directly to a sale. Moreover, sales figures are also distorted by prevailing economic conditions, such as a recession, and in such cases an effective advertising campaign may be associated with falling sales revenues. Nonetheless, it is clearly possible and desirable to make some efforts to establish whether or not a new sales enquiry has resulted from a particular advertisement or advertising media.

Most major campaigns are designed and mounted by advertising agencies, who will also arrange for tracking studies to be carried out. These tracking studies will attempt to ascertain both the proportion of the target market that can recall seeing a particular advertisement, and the lasting message that has resulted from it. Clearly, the former is easier to assess and the success of an advertisement is normally judged in terms of recall, but it is worth noting that the criterion can be misleading if the advertising message does not motivate desire or action on the part of the target. The long running series of Martini advertisements, featuring Joan Collins and Leonard Rossiter, was recalled and enjoyed by vast numbers of people, but the positive brand images in these adverts were generally undermined by the punch line – which required the drink to be thrown over the Joan Collins character. Consequently, insufficient numbers of young people were motivated to try the products, leading to a decline in sales over the longer term.

Money

An overall budget for advertising must be set. In planning terms, this would be done by setting realistic goals for sales figures, and then estimating the advertising support needed to achieve these targets. In practice however, the advertising budget may be set in an arbitrary fashion, based on last year's sales revenue, or based on what the company accountant thinks can be afforded. Sometimes advertising budgets are set by reference to a competitor's level of activity. Advertising costs are typically around 2 per cent of total costs.

The five Ms and the sales force activity

Since engineers are increasingly involved in sales, as both support staff and sales engineers in their own right, it is worth outlining the importance of advertising and promotion in salesforce activity, and the promotional aspects of the sales representative's role.

Money

A small company may not be able to maintain its own salesforce, and may have to rely on agents to carry out the sales function. A larger firm will have to set budgets for the whole salesforce, broken down into regions (if the salesforce is so organized), and for individual sales representatives.

These budgets may be set according to a variety of means, ideally with reference to sales targets, and typically account for around 3 to 5 per cent of total costs.

Motion

Motion here relates to the frequency and objectives of sales visits. Selling requires mastery of several distinct skills; including the ability to interest, create empathy with, and motivate potential buyers, before ultimately negotiating and gaining a final agreement (commonly called closing a deal). The initial stages of motivation may require several visits to help buyers make up their minds. The salesperson will have to time these follow-up visits skilfully; too soon and a buyer may feel pressured; too long and the lead will go cold. Moreover, such visits are ideally co-ordinated with other visits in the area, to stay within budget.

Sales and promotional literature is of crucial importance here. It provides useful discussion topics to justify further visits to both major distributors and customers, and new prospective distributors and customers, on a regular and phased basis.

Message

Sales visits are a media to authenticate the advertising and sales promotional literature, to an interested audience of customers, in the case of direct selling, and to distributors, retailers and end users in the case of channel marketing.

It is vital that both product information (particularly regarding new products), and a positive corporate image are presented at a sales interview. This is particularly true when reinforcing performance track record to a major customer at the time of renewing a contract. The fear is that the customer, although not really wanting to change suppliers, is also aware of the danger of complacency among incumbent suppliers, and may wish to take the opportunity to review alternatives. The existing supplier may have an advantage over competitors if the salesperson has built up strong rapport with the purchaser, and is aware of the kind of information and means of presenting the full value of the offering in a way most likely to influence the decision favourably.

Media

Sales staff commonly meet clients at the client's office for an interview. New products, or significant modifications are often launched at sales seminars (also called sales conferences and new product seminars), where clients and potential clients are invited to see sales presentations and demonstrations.

Exhibitions, trade shows and fairs have been useful selling aids in helping customers make up their minds to make decisions regarding which products are of value to them. But there is now a significant move away from deploying sales staff at these events, replacing them with Public Relations (PR) staff. Partly, this is because such media generate relatively few sales leads, consequently sales staff are better deployed elsewhere. However, it is also worth noting that the message recently given at shows and exhibitions for many industries has been concerned with proclaiming a firm's continued survival, rather than an aggressive sales pitch.

Measurement

The effectiveness of individual sales staff can be assessed in a large number of ways, including the number of new clients found, the number of contracts retained or lost to competitors, the number of sales calls made, the degree to which price has been discounted to make a deal, the number of new products sold, the timely submission of quotes to customers and the number of monthly sales reports submitted on time. Good assessment tries to take into account the regional or market factors which may have influenced a particular salesperson's performance.

Assessing the performance of the sales team as a whole is rather more problematical. As always with promotional activity, it is dangerous to seek a direct correlation between the size of the budget, and extra revenue generated as a direct result of increased expenditure. A key topic at monthly sales meetings is the use of promotional and advertising material to support sales tactics. A bigger promotional spend may increase turnover with established customers, through enhancing the augmented offering, and help retain custom from wavering clients, although the extent to which promotional material can offset poor quality in anything but the short term is limited. But for new accounts, the strategy of breaking in against established competition is very much a credibility battle, where the prospective customer will test all aspects of claims made by a sales representative, before giving a trial.

It is winning this trial which is the objective of the sales and promotional campaign, and targets should be set accordingly. Increased sales may or may not result, depending on how well the trial product performs, the price and delivery times quoted, or any other part of the marketing mix. It would be most unwise to incorporate these into salesforce efficiency measurement, indirectly, by adoption of unsuitable criteria.

In this section we have examined the objectives and techniques used by marketers to communicate the value which has been created by earlier stages in the value creation process. But it should be clear that the promotions mix is most effective when used to support genuine innovation and quality created during these earlier stages, and will have little sustained impact if this is not the case.

Chapter summary

In Chapter 2 we have treated business activity as a process of value creation, and argued that, in a marketing orientated company, this process has three stages: identification, creation and communication of value. The engineer has a vital role to play in the second stage, but long-term prosperity requires that these engineering skills be deployed within the context of market orientation.

We have also briefly examined the marketing tactics and techniques open to the marketer. The marketing mix involves four elements; product, price, promotion and distribution, which must be combined to deliver the correct product offering to the right market segment, supported by an appropriate promotional mix. Engineers are to be found contributing to all four. Further, these engineering and marketing skills must be combined effectively in order to create sustainable competitive advantage by cost leadership, differentiation or focus strategies.

Further reading

There are many excellent introductions to marketing which take the issues and problems further than we have attempted here, including: *Elements of Marketing*, by A. Morden (DPP, 1987), *The Fundamentals and Practice of Marketing*, by J. Wilmhurst (Heinemann, 1984), and *The Marketing Book*, a useful collection of reading edited by M. Baker (Butterworth-Heinemann, 1991). A recent work which helpfully directs marketing activity towards achieving competitive advantage is *Competitive Positioning*, by G. Hooley and J. Saunders (Prentice Hall, 1993).

Tutorial exercises

Case study 2.2: Xylogics International

A modern computer network can have hundreds, or even thousands, of computers, terminals, printers, modems, disc drives and other peripherals linked together, as well as ports that interact with other networks. There are several types of network operating systems available, UNIX being one of the most popular.

The device, and the associated software which enables system users to access and use parts of the network, is called a communications server. The fast growing market for communications servers was estimated to be worth around £250 million in 1993, with key markets in the UK, USA, Europe and Pacific Rim.

Xylogics are a major competitor in this market, with a 22 per cent share of the ubiquitous UNIX market, and a major presence in other markets. The American and Pacific Rim markets are run from the USA, while Xylogics International, based at Milton Keynes in the UK, runs the rest of the operation.

Market segmentation

Xylogics International segment each international market by types of buyer. First, there are major companies designing and manufacturing major network systems, such as ICL and NCR, who will customize Xylogics' communication servers, incorporating them into their own range. This market is referred to as the Original Equipment Manufacturers (OEM) market. Second, there are buyers who use communication servers in the design, installation, modification and upgrading of networks. These are known as value added retailers (VAR).

In both segments, the customer, or buyer, will not be the end user. The end user will probably not be aware of Xylogics' name, or its ANNEX brand products. However, the needs of both buyers and end user must be considered when assembling the product mix.

Price and products

With large network systems, the communications server is a tiny fraction of the network cost. Buyer decisions will be based on quality, reliability and product features, so Xylogics base their strategy on offering good value for money, rather than low price; offering a wide variety of features, such as flexibility, tight and invisible integration with the UNIX system (invisible in the sense that the user will be unaware of the ANNEX system, as it requires no extra training or commands to operate), centralized network management, modem and printer support and added security against hackers and illegal users.

Moreover, there are a range of options offered with each product in the range.

Small network users tend to be much more price conscious, although a substantial segment require advanced features at low cost. Accordingly, the company produce a basic product, which aims to be the lowest priced on the market, and an advanced model, with a greater number of enhanced features, which operates on a value for money basis.

Distribution

Since the OEM segment is different to the network user, the products, although identical, must be distributed differently.

In the OEM market, one contract may be worth several hundred thousand pounds, and buyers will want to negotiate individually. Therefore, there is the need, volume and margin to support a specialist sales team, selling direct to OEMs, operating from sales subsidiaries in the UK and Germany.

In the end-user market, individual contracts are generally worth much less, and so a two-tier system is used. Each country has one high level distributor, who, in turn, sells on to other resellers of communications servers, and acts as a point of contact and support. The high level distributors are not agents, they buy the communications servers before they are sold on (as opposed to an agent, who would merely arrange a deal between Xylogics and a reseller, and take a commission). Xylogics make 2 per cent of the invoiced revenue available to the distributor, through a marketing co-operative fund, to support reseller promotional activity. Recently, the French distributor used this fund to hold a series of sales seminars to promote a new product launch.

Low level distributors include consultants, systems analysts and integrators, networking and UNIX resellers, application software houses and UNIX resellers.

Xylogics have watched competitors, and manufacturers of PCs and other hardware, as well as software writers for some other information technology hardware, dispense with intermediaries and deal direct. However, Xylogics believe their product to be sufficiently distinctive, and carry a margin adequate to support the longer channel.

Promotions

Major promotional objectives include:

- To broadcast the continued existence and dynamism of the company
- To reinforce the purchasing decision
- To support distributor advertisement to end-users in each country

- To advertise new product launches and new product features
- To generate new sales leads
- To contact more VARs

These objectives to be achieved with the following media:

- Advertising in appropriate specialist IT magazines, and the VAR Seller, Integration and Networks
- Presence at Information Technology trade conferences and exhibitions
- A quarterly newsletter, press releases and technical bulletins, sent free to resellers

All subsequent enquiries from end-users are passed to appropriate VARs.

Recent trends

Xylogics are now placing greater emphasis on finding new VARs to sell and support the product range. One recent project included a programme of 20,000 inserts into a variety of publications, which returned a low, but acceptable half per cent response. Far more successful was a telemarketing campaign, carried out by a professional telemarketing agency, targeted at 500 leading VARs in the UK, which produced a 10 per cent response.

Promotions effectiveness

Xylogics do not routinely monitor their campaigns for effectiveness and cost efficiency, although they do have expectations for much of their activity at the design stage. However, many of their objectives are not quantifiable and, moreover, their advertising message is aimed more at 'Selling a concept rather than a product, and creating brand awareness'.

2.1 (a) How would you match Xylogics' advertising message to the media available? Give examples of the kind of copy which would convey a suitable message.

(b) What possibilities are open to distributors for using the co-operative fund?

(c) Would it be possible to develop more stringent criteria for assessing Xylogics' promotions expenditure? Would there be any great value in doing so?

(d) Xylogics is a very successful producer of one product range in a fast growing market. Presumably, at some (hopefully) far distant time, the market will slow down, and may decline. What will happen to Xylogics? Is there anything that can be done in the meantime?

2.2 Why should distributors disappear from such markets as the PC market?

Case study 2.3: Notts Knitting

Notts Knitting Ltd is a private limited company founded by a qualified engineer, Harry Mills, who is now 67 years old and chairman of the company. The company was formed some twenty years ago, to manufacture knitting machines for the textiles industry. In the mid-1970s, the emphasis was changed somewhat, in the face of European, particularly Italian, competition. The company has focused on developing specialist machines for the trickier parts of the process, including stitching and buttoning, which it will customize on request. By 1980, the company had developed a salesforce across the UK and occasionally sold in Europe.

The UK industry consists of around ten major producers and a number of smaller, local firms. Notts Knitting has grown to be one of the top ten, in sales volume terms, because of its highly competitive pricing, according to the sales manager. Recently, competitors have become extremely vigorous, especially in new product innovation. But Notts' engineers have so far succeeded in placing a close substitute on the market soon after each innovation.

The sales staff are based in three locations, Manchester, Leicester and Bristol, while the production plant remains on its original site in Nottingham. The salesmen and women are extremely active and effective in terms of contacts, customer relations, dispersion of product information and obtaining sales.

The engineering department is manned by a staff of very competent engineers, well versed in the various technical areas needed to design and produce their specialist machines, though at present there is limited use for new technology in design and manufacture.

As of late, the production manager has received numerous complaints from the engineering department regarding the methods of the sales staff. It seems that individual sales representatives contact the engineering department's designers whenever they find a competitor has introduced a new variation to the product, or a customer requests a new customization. The design engineers then supply a preliminary design and cost estimates to the sales team, so that specific details on size, cost and performance and so on may be shown to any potential buyer. Over the last eighteen months, the system has broken down somewhat, a recent incident being an order for a radically different type of machine to attach stiff collars to formal shirts. The engineers had previously provided preliminary designs and cost estimates, as normal. But the specifications for the final order were completely different to those of the preliminary design.

The engineers' complaints were voiced by the chief engineer at a recent meeting of the board of directors, who said, 'The sale staff come in here and ask us for preliminary designs and we break our backs to provide them at very short notice. Then they come back and tell us they've sold a completely different machine. Of course, we have done no preliminary work, and we are behind from the start. It seems they are prepared to sell any old thing the customer wants, rather than the quality products we design for them.'

Once started, the chief engineer continued with a catalogue of other complaints, including the practice of negotiating individual prices with each customer.

'Hasn't anyone heard of standard pricing?' he exclaimed, looking directly at the sales director, 'and your people seem to sell whatever they want, whenever they want. Some months we stand around idle waiting for business to come in, and other times we are rushed off our feet. We have to go to excessive overtime, and even then we let down half the customers who've been given impossible promises. They all come on the phone to us, and we end up a damn PR department. It's time you sorted your lot out.'

The sales director replied in a short memo:

'The success of this company is entirely due to the skill of my sales staff. I have complete faith in their organization and methods, so much so that far from, "sort them out", I see no need for central control, visits or sales conferences. And if the engineers can't keep up with us, I suggest we sub-contract the work out to more flexible suppliers.

'The engineers are always moaning about something. But we could get over this by better understanding and improved communications. I will inform my sales people to channel all their new sales inquiries through the chief engineer. OK.?'

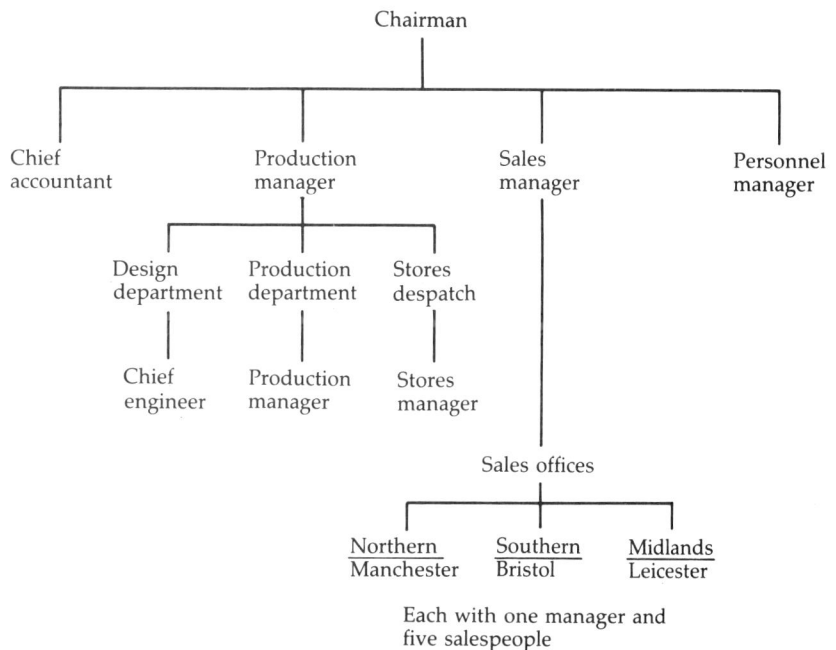

Figure 2.4
Organization structure of Notts Knitting

2.3 (a) What do you regard as being the main source of conflict between engineering and sales at Notts. Knitting? What steps could you take to resolve the problems?

(b) Could Notts Knitting be transformed into a marketing-led company? How would this affect the work of the engineering department?

2.4 *Segmentation exercise*

Are all men consumers the same? Are all women consumers the same? Attempt to segment men and women into various personality types, and examine how each type might respond to:

- tinned exotic fruit salad
- a very expensive, but virtually silent power drill
- an environmentally friendly chocolate bar
- the BMW hydrogen powered motor-car discussed on page 14

Note: The following classifications have been developed by the advertising agency McCann and Erickson, and may help in identifying other 'types' of men and women.

Women

Avant Guardian, 10 per cent of women
Liberal opinions, trendy concerns and attitudes, outgoing, sociable, active.

Lady Righteous, 16 per cent of women
Traditional, conservative views, happy/complacent, strong family orientation

Downtrodden, 13 per cent of women
Shy, put upon and easily manipulated. Unhappy and pressurized in relationships

Men

Pleasure Orientated Man, 9 per cent of men
Has a self image of masculinity and leadership. Often self centred, disliking work and easily tempted into impulse purchases.

The Traditionalist, 16 per cent of men
Emphasizes respectability and 'correct' attitudes, rules and behaviour. Values security and self respect, and is generally conservative.

Sophisticated Man, 10 per cent of men
Intellectually motivated with broad interests and social concerns. Admires artistic and intellectual achievements, and is attracted by the new and fashionable.

3 Applying the marketing concepts

The strategic planning process

In the previous chapter, we looked at how firms identify, make and promote value in a way profitable to the organization. In this chapter we examine the means by which an organization ensures its survival by focusing its activities toward markets which will become profitable for the organization in the longer term. The term strategic is used for several reasons, most importantly to imply that the plans are based on the future, and second, to distinguish it from operations (things that people actually do) which, although different, should be based on this long-term view; rather than short-term expediency.

This orientation toward the longer term is achieved by the strategic planning process, which has five broad stages, each designed to answer a particular question, as outlined below:

The five questions of strategic management

1 *Where are we now*? This is an audit of the internal and external environments. The internal environment includes those things within the organization which are controllable by the organization. The external organization is the world at large, in which the organization is usually much more reactive. This stage also involves definition of core business(es) into groups called 'strategic business units', (SBUs). SBUs are recognizable business entities focused at well identified market sectors, for example, a financial services company might think of its insurance business as one SBU, its chain of estate agents as another. The health and prospects for each SBU would be assessed individually, by the process described under 'Strategic Health Check' below.

2 *Where are we going*? This involves analysis of the alternatives available to the organization, in terms of each SBU. The resources and potential identified in the auditing at the first stage help define the broad direction(s) in which the organization should be heading in future. To be effective, it should take account of the internal and external environmental changes occurring continuously. At this stage, top management sets clear missions for the organization in the form of a long-term mission statement; designed to stretch the whole organization well beyond its present performance. Mission statements should be short and believable, so that consistent operational decisions and objectives can be devised.

3 *How are we going to get there*? This involves narrowing down the broad strategic picture into tangible marketing plans. Product development strategies and manufacturing investment implications would be integral parts of the planning process, and critically, the resources necessary to achieve these plans in terms of training needed, production capacity and finance, should

be identified in sufficient detail for meaningful delegation. At this stage, some decisions should be made about competitive strategy, to determine whether or not the SBU's strengths and weaknesses, together with the market conditions faced, suggest a cost leadership, differentiation or focus strategy as the best option.

4 *What are we going to do*? Nothing happens without people, be it individually, or as teams, being clear on exactly what they are accountable for. At this stage, operational plans are set in terms which enable management to organize, delegate and to do the tasks necessary to achieve the plan.

5 *How will we know when we've got there*? The organization must regularly monitor operations, in order to check whether or not the firm's objectives are going to be achieved, whether contingency plans need be invoked or objectives need to be reset in the light of changes in the business environment.

The five stages of strategic management

Each question is answered by a stage in the strategic planning process. These five stages form the structure for the remainder of this chapter.

Stage 1: The strategic audit

This part of the process aims to answer the 'Where are we now?' question. The purpose of the audit is to establish, as far as possible, answers to the following questions:

- What markets is the firm in?
- Who are the firm's customers in those markets?
- What value is created for, and appreciated by these markets and their respective customers?
- Are the strengths of the firm particularly suited to this business?
- Are trends in the wider business environment favourable to the firm's activities?
- Will unfavourable trends in the business environment exacerbate the effects of the firm's weaknesses?
- Who is the competition? And what are they doing that will make our current mission more difficult to achieve?

Auditing the external business environment

An analysis of external business issues consists of sub-dividing the complex business environment into four sub-sections; namely political, economic, social and technological factors (PEST). PEST analysis (also called STEP) is a vital tool for structuring external intelligence, as it is a review of all the beneficial and problematic external issues which affect a business. There are no textbooks for this approach: it consists of being aware of current affairs, and that means news articles from all sorts of media, increasingly global in nature. Important sources include the *Financial Times*, CNN News, BBC 2's *Newsnight* and *Money* programmes, *Business Week* magazine, and the host of banking weeklies and monthlies on economic intelligence available in major libraries. In the case of engineering products, technical journals will be essential in ensuring that the implications of new and developing technology can be anticipated and accommodated in the firm's strategic plans. This bespoke research is done by the research department, usually within the marketing intelligence or corporate strategy department in the case of big businesses who can afford this continuous industry watching; otherwise this is the bread and butter of many independent consultancy firms.

PEST might imply that issues fall neatly into one category, but in reality, major issues spread themselves. For example, the re-unification of Germany is clearly a political event, yet its long-term effects on the Exchange Rate Mechanism and financial stability within Europe have profound economic effects, and the growing number of economic refugees will undoubtedly have social effects. The important thing is to ensure that all possible effects are considered, rather than debate whether an event is political or social.

The aim of PEST is to produce a series of possible or actual events in each environment in the form of bullet headings, which would then be evaluated by corporate strategists. Below, we give examples of the kind of events which would require serious consideration.

Political

This element of the environment includes political events and legal systems. Important events would include, for example, effects of the Single Market in Europe, tightening of product liability laws by the EC. Examples include pressure on the British Water Authorities to improve the quality of tap water; forthcoming emission laws on the car industry, instability and strife: such as trucking companies having their perishable produce stuck on French motorways in July 1992 as a result of a transport strike in that country. Unfortunately, it is not possible accurately to forecast political events which may have a catastrophic effect on business; for example the success of the oil exporting countries in raising oil prices in the 1970s had immediate profound effects on prosperity and growth in the West, and catalyzed product design for energy efficient products.

Economic

This area would include, in a wide sense, the demographic details of population ages, birth rates and death rates; economic details of income and income growth, inflation, unemployment levels, economic growth and taxes; policy details of government spending, intervention in industry, trade barriers, exchange rate management etc. In many ways, this particular environment is becoming increasingly difficult to manage strategically, since instant, ephemeral events continually upset longer term trends. Indeed, some writers have questioned the whole idea of long-term strategic management, on the basis that the economic environment is now too volatile to forecast and plan for. A sneeze in the Tokyo Nikkei or Hong Kong's Hang Seng is felt in Wall Street, and then reverberates around Europe. Similarly, continual political upheaval in Eastern Europe acts as a barrier to economic commitment from the IMF (International Monetary Fund). More locally, uncertainty has increased at business level, when forecasting say, export revenues accurately because of fluctuating exchange rates, since sterling's unglamorous exit from the Exchange Rate Mechanism in September 1992.

At a closer level, the organization will be interested in analysing data about customers, market segments and financial backers. In particular, a firm will need to undertake a competitive analysis, based around the following five forces of competition:

- the strengths and weaknesses of its competitors
- the relative bargaining power of suppliers
- the relative strengths of customers
- the threats of further entry into the market by companies producing a similar product
- the threat posed by substitute products which can satisfy similar needs in an alternative way.

To restate the obvious, once a company has identified threatening competitive forces, it must develop a competitive strategy based around developing cost leadership, differentiation or focus strategy.

Social

If the power of social forces is ignored at political and industrial level, perilous situations can develop or great opportunities can be missed. The huge attention to the ozone layer led to steps virtually to outlaw the use of CFCs in aerosol canisters, and drive the refrigeration industry to seek more costly alternatives to the environmentally unfriendly freon. Similar pressure has led to the banning of ivory hunting and whaling, although contrary pressures remain. Even if legislative support is not forthcoming or practical, there remain marketing opportunities to provide environmentally friendlier products. Health and Safety laws now reinforce social pressure in favour of the consumer, i.e. the end user of most products, who is protected from product defects and injury by laws which are now EC wide, following in the footsteps of Ralph Nader's campaign in the USA in the 1970s.

Even without legal reinforcement, many companies are now talking explicitly about ethical decisions, for example, IBM have internal rules which preclude some legal production methods because the company judge the resultant environmental impact to be unacceptable. Ethics no doubt limit the scope for the business organization, although may have some pay-off in terms of corporate identity and promotions.

Social trends would also include changes in behaviour and life-style. For example, during the 1980s the habit of house buying spread to many social groups who had formerly lived in public sector rented accommodation. This created many opportunities for selling services associated with house purchasing, and boosted DIY and domestic construction markets. The poor performance of the housing market in the late 1980s and early 1990s has caused enormous pressure on suppliers in those markets, regardless of how well their products and services were formally marketed.

Technological

This environment includes changes in the technology embodied within the existing products, technology which enables a well known need to be satisfied in a new way and new technology which enables new needs to be identified and satisfied. Technology used within the production process itself will be reviewed continually.

The constant pressure to develop and maintain a competitive edge through innovation means that companies must always be alive to alternative and substitute methods of providing added value. For instance, the compact disc has rendered cellulose records largely obsolete. Many firms now find that products and services need to be redesigned and updated ever more quickly, and have to strive for innovative excellence, just to remain in the competitive arena. This innovation applies equally to design originality and new manufacturing methods which add value to products, or provide better quality at the same value.

PEST analysis is focused on events in the external environment, that is, on events and trends over which the organization had no, or little, direct control. It is a vital input to the next stage of the audit, which looks more closely at the firm's performance and potential.

Case study 3.1: The engineering environment

The engineering employers' update of its spring report on economic trends gave little cause for optimism in the engineering industry, which accounts for between 8 and 9 per cent of GDP. Engineering output is now stable, but the expected upturn has been postponed until late this year.

The EEF forecast of engineering output growth in the fifteen months to mid-1993 has been revised from 4.5 to 2.9 per cent.

The federation blames the international economic climate and delayed British recovery for the need to scale down its forecasts. Before the general election it predicted a 0.5 per cent growth in 1992, but now expects a 0.6 per cent decline.

Engineering output has fallen by 17 per cent in two years but the report predicts a 4 per cent recovery over the next year. Imports are expected to rise by 6 per cent and exports by 4 per cent over the same period. The report predicts the upturn in demand will come from deferred investment, particularly in the motor industry.

In 1992 prices output from British engineering industry at £126m is lower than it was 12 years ago. If the federation's forecasts for next year are correct, 1993 output will still only achieve the level of 1980 – though productivity has increased massively over the same period. Engineering employment has fallen by 260,000 to 1.8 million in the past two years and is expected to drop by another 60,000.

The Guardian, 24 July 1992

Internal audit

This part of the audit examines how the organization's resources are used. The aim is to produce a summary of the organization's strengths and weaknesses.

It is most common to start the analysis at departmental level, since management functions tend to be so grouped, although there is no technical reason why this must be the case. The kind of issues raised in such an audit are given below:

Marketing: Are products and brands strong, or past their sell-by date? Are distribution channels effective and competitive etc.? How good are the sales people, are there enough of them and are they properly motivated?

One particularly helpful approach is through the eighty:twenty rule, also called Pareto's rule, which would suggest that around 80 per cent of the company's profit will be made by 20 per cent of its products. Although imprecise when stated in this way, identification and analysis of profit/product ranges, and many other variables, do imply that the eighty:twenty distribution does approximate the real world surprisingly well. The purpose of using it in the audit is to identify and safeguard the important 20 per cent or so of the organization's core activities.

Engineering: What design, R&D and engineering skills are available, and how good are they? How flexible is the engineering department, what is the time needed for customer specifications to trigger a working design, compared with competitors?

Manufacturing: What technology and systems are used, how does this affect quality, flexibility and cost compared with competitors? How long is the lag between a working design and a delivered product? Pareto's rule has implications, particularly with regard to manufacturing cost control. For example, around 80 per cent of component costs come from around 20 per cent of parts. Tight inventory control of that 20 per cent would help drive down costs. Similar relationships can be found in scheduling throughput and other areas.

Finance: Does the firm have sufficient reserves of capital, could more capital be raised, is the firm prudent in the management of cash on a day-to-day basis? Is the firm making sufficient profit to justify its continued existence, and support further re-investment? Again, the eighty:twenty rule has implications for cost control, particularly with regard to debtors.

HRM: What skills does the organization have; and are these sufficient to enable further development and growth? How does the organization respond to change and stress, is it unnecessarily bureaucratic or conservative, or is the culture innovative and adventurous?

The issue here is largely of efficiency, that is, 'How well do we do things'? However, the audit will also look at effectiveness, which questions the purpose of such activity. Effectiveness is particularly concerned with value creation and competitiveness (see Ghastly Gadgets case).

Case study 3.2: Efficiency and effectiveness

Ghastly Gadgets

Ghastly Gadgets have made cheap, low technology components for several generations. Recent growth in the DIY market has allowed the firm to expand considerably. A few years ago, the firm thought that it was sufficiently large to run its own transport fleet, which it did by purchasing a fleet of seven and a half ton trucks. This fleet is now due for replacement, and the following questions have been raised.

Efficiency questions: Are seven and a half ton trucks the best size for our needs?

Is the former supplier still offering the best deal?

Should we buy or lease the vehicles?

Effectiveness: Why are we running a *trucking* operation? Does having our own transport fleet add to competitiveness, and if so, how?

If the management of this fleet is so good that the service offered is particularly flexible, low cost or distinctive in some other way, perhaps the organization should consider going into the trucking business. On the other hand, if there is nothing distinctive and no value added, the firm should reconsider its decision to run its own fleet, and use an outside contractor. In this way, managerial energy, professional expertise and financial resources could be focused on things where it has particular competence and can add value and competitiveness.

SWOT analysis

SWOT stands for Strengths, Weaknesses, Opportunities and Threats, and is the means by which internal and external audits are pulled together coherently. Strengths and Weaknesses are best analysed by looking at the performance (i.e. efficiency and effectiveness) of managerial functions and resources, as described in the preceding section. The PEST analysis, also discussed above, will identify the outside threats and opportunities developing in the business environment that the organization is operating in.

SWOT is useful in that it enables these opportunities and threats to be matched to the strengths and weaknesses. By way of example, a SWOT analysis on the case study 'Story of a Sole Trader' from page 13 is reproduced below.

Smith's strengths were primarily technical in nature, and this would help him exploit the opportunity he had identified. However, the way in which he set up his business actually compounded his weaknesses to create major problems later. In particular, his decision to set up his own workshop seems questionable since he had limited non-technical business experience, and a limited capital base. He could have built a business on his strengths by selling his designs or design consultancy. If he had wanted to move into production, forming some kind of association with one of his engineering contacts (either a partnership, licensing agreement, leasing of surplus premises and equipment or simply a contract to manufacture, perhaps under Smith's supervision) would have reduced the need to commit all his capital to the project and enabled him to specialize in exploiting his strengths.

By the time Smith had identified a second, much bigger market for his products, his weaknesses had multiplied the problems.

SWOT helps the analyst identify the important factors in a given situation.

Strengths Smith's skills in design. Smith's skills in production. Well-made, well-designed products. Support of banker. Large number of contacts among local · engineering companies. Free workshop.	**Weaknesses** Smith's limited managerial skills and experience. No selling experience. Capital base limited by the value of Smith's house. Small size of the workshop. Limited capacity. Limited market research.
Opportunities Large local demand for Smith's automotive products.	**Threats** Growing unrest in steel industry (possibility of steel shortages in future). Penalties in workshop lease. Possible recession in automotive markets. Growing automotive imports.

Figure 3.1 Smith at the beginning of the enterprise.

Strengths Smith's design skills. Smith's production skills. Smith's selling skills. Well-made, well-designed products. Free workshop.	**Weaknesses** Insufficient plant to meet order. Workshop too small. No capital reserves. Large and growing overdraft. Insufficient money to pay trade creditors. Cash lost to bad (and doubtful) trade debtors. Hesitant banker. Wife's income also dependent on business.
Opportunities Selling to large-scale automotive engineering companies. Non-automotive engineering?	**Threats** Depressed local demand for engineering products. Continued steel shortages. Increasing pressure from creditors. Penalties in workshop lease. Continued problems in automotive industry.

Figure 3.2 Smith at the end of the enterprise

The SWOT analysis is a vital first prerequisite of Stage 2, as it highlights feasible and, equally, non-feasible options for the organization to consider. There are four combinations of SWOT factors; the following depicts typical reasoning in a SWOT analysis:

1 A strength can be married to an opportunity
2 A weakness can be married to an opportunity
3 A threat can be married to a weakness
4 A threat can be married to strength

1 If strength can be married to an opportunity, then this would minimize the risk of failure, as the organization has strengths compatible with the identified opportunity. Peter Smith's original business idea was one such example, a large local demand for products which he was able to make with his distinctive competencies.

Oddly enough, such an situation can arise through a potential threat, for example, the move towards replacing CFC-based propellants in aerosols is a threat to existing product ranges of all such producers, but those who have developed alternatives already may consider this a strength and opportunity situation, since competitors will be forced to catch up with costly R&D programmes. No doubt the BMW prototype of the hydrogen powered car discussed in Chapter 1 owes much to this kind of long-range thinking.

2 If an opportunity arises, but is sufficiently removed from the organization's strengths to cause concern, stratagems must be employed to overcome, or at least diminish the effects of these weaknesses. Peter Smith's decision to produce his products with his own enterprise is, in fact, an example of this opportunity/weakness condition. Alternatives, such as selling the designs, consultancy, factoring, leasing workshop space from his business acquaintances or a full partnership may well have offset his own inexperience and limited capital and organizational problems. In general, managers should think carefully before straying too far from the organization's main strengths and competencies, since this tends to undermine competitiveness. But where an opportunity is too big to be ignored, such as large growth of a particular segment, the organization's strategy would be inward looking; aiming to restructure so that these weaknesses can be at least overcome, and hopefully turned into strengths. Such a transformation takes in the change management process; the subject of Chapter 12.

3 A threat can arise which hits a firm particularly hard because it strikes at a weakness. This can happen if the company is unprepared, or carries out its business in a fashion which makes it more vulnerable than its competitors. The difference between an external threat and a weakness, is one of control. A weakness is an internal problem which management have some (although often limited) power to control, while a threat is a PEST factor which the firm can react to, but cannot control.

Where there is no internal power to control, action plans must be made to contain that threat if the scale of the threat is large, and the ability of the organization to cope with the threat is small. Clearly, an inability to cope with certain types of threat is a weakness, and should be identified as such during the internal audit. For example, many businesses are currently going into liquidation because they are under capitalized, meaning that the business is run with insufficient funds. Such firms can make profits – even large profits – but may become vulnerable if a major customer defaults, if the cost of borrowing rises or there is a general move toward sales on, say, 30 days credit rather than cash.

In terms of Peter Smith's case, the steel strike was such an external problem, but it is probable that his self-induced weaknesses caused much of the resulting trouble. All similar firms would have had difficulty in obtaining supplies of steel, but in his case the financial problems, poor control of credit to customers, inattention to cash flow, and undercapitalization actually brought about his downfall. Many of the suggestions listed in 2. above would have eased the problems, but factoring of invoices, leasing rather than buying his equipment and developing relationships with the larger firms in the industry might have eased the situation.

47

4 A threat may, somewhat perversely, be linked to an internal strength. For example, a strong design team is undoubtedly a strength, but the organization will have to ensure that the team is not head-hunted by a rival firm. Creative teams in advertising agencies and money market speculators can often be head-hunted *en masse*.

Conversely, an external threat which is linked to the firm's strengths may often be turned to an opportunity. By definition, strength implies that the organization is likely to respond better than its competitors and, in the longer term, may find the competitive position improving as weaker firms fall victim to the same threat, as discussed in 1. above. However, this is by no means guaranteed; particularly if the impact of the threat is long term. In such cases, a major strategic review may be necessary to reset the mission of the organization.

In reality, of course, threats and opportunities do not readily fall into nice strength and weakness categories. Each situation must be analysed in terms of the organization's future strategic priorities, and the effect it may have on competitive strategy and current operations. For example, the analysis of Peter Smith's business clearly reveals that his activities throughout the story, and not just at the end, were determined more by his weaknesses than his strengths, and the causes of his ultimate failure were present from the beginning. SWOT helps us to see that the major cause of Smith's problems was his own poor decisions, rather than unfortunate events beyond his control. The case demonstrates that he was as unprepared for success as for failure.

The idea of a strategic business unit

The 'Story of a sole trader' illustrates the way to base strategic analysis for a particular business. It is made simple by the fact that the whole business is owned and managed by one person, on a single site, and all the activity is aimed at one particular market. For many engineering companies, this is not the case; the organization may be multinational, dealing in many related and unrelated markets. For example, Rolls-Royce have interests in aeroplane engines, gas turbines and power generation (see case study on page 161). Moreover, companies like Hanson and Lonro own a great diversity of enterprises in many industries and countries. A SWOT analysis for the whole of a diverse, multi-industry company would be difficult and pointless. It is more common to think of such companies as being composed of several individual Strategic Business Units (SBUs). Each SBU should provide a separate product range to a distinct market segment, and have identifiable competitors.

Once the SBUs have been identified individually, it is sensible to develop SWOT and PEST analysis for each of them. However, a further type of analysis also becomes possible, the strategic health check.

Case study 3.3: Defining SBUs

Bethlehem splits plants into separate businesses

Bethlehem Steel, the second-largest US steel manufacturer, is splitting its two large plants into separate business units in order to make them more financially accountable and bring them closer to their customers.

The move is a significant departure for a company which has the reputation as one of the more conservative of America's integrated steel manufacturers. It is one of the first initiatives by the company's new chairman, Mr Curtis Burnette, who was formerly Bethlehem Steel's vice-president. Mr Burnette took over last year on the retirement of Mr Walter Williams.

The new units are based at the company's two flat-rolled steel plants at Burns Harbour, Indiana, and Sparrows Point, Maryland. The two plants together produce some 7.5m tons of steel a year. Each will now be responsible for its own marketing, operations and financial performance.

As part of the move, Bethlehem's product marketing operations, run from group headquarters in Bethlehem, Pennsylvania, will be split between the two units. The move will make Bethlehem's structure similar to that of low cost mini-mills.

The mini-mills, which make steel from scrap metal, have made substantial inroads, over the last twenty years, into markets once dominated by the integrated manufacturers. Mr Burnette said 'major benefits of the business unit structure include improved customer focus, responsiveness, speed of decision-making, employee commitment and business awareness. It places the responsibility for the success of the business in the hands of those at the decision level.'

Financial Times, 20 January 1993

The strategic health check

Once SBUs have been identified and associated with a particular market segment, it is possible to make some inferences about the long-term prospects of that SBU, and place its marketing and financing activities into a strategic perspective.

For example, there has long been a need to transmit urgent business information. Consequently, engineers of various types have devoted much attention to devising more efficient systems; including telegrams, telephones, telexes, fax and E-mail systems. As each system is invented, it reduces the growth opportunities of (or even completely replaces), its predecessor, and, in turn, is eventually superseded by a technology which produces greater value, or the same value more cheaply. This cycle has been observed in a great many industries over a long period, and is referred to as the product life cycle.

Suppose a large company had several SBUs, each producing a different type of communications product (i.e. it might have had an SBU making telephones, another making portable two-way radios etc.). One SBU, called TELL X, set up to produce better and cheaper telex machines, would have found itself in great difficulties as fax machines began to dominate the market for business information transmission systems. Some fundamental realignment of the SBU's activities would have been a matter of priority; clearly TELL X would have to adapt to the new technology or go out of business. A SWOT analysis of TELL X would show whether or not it would have been worth realigning the SBU to design and produce fax machines, or whether it would have been better to scrap or sell the SBU and acquire an established fax producer. The similarities and differences between the technology required in telex and that for fax would play a major part in the decision.

It is important to see that the decision to stop developing telex machines is a strategic one; that is, it is based on long-term market movements, rather than the operational success or failure of TELL X's current performance.

Business strategists have developed a wide variety of analytical techniques and models to assist in a strategic health check (also referred to as a strategic audit). Below, we discuss one such model, the product life cycle, selected for its simplicity, robustness over time and marketing orientation.

Product life cycle

The product life cycle is a stylized representation of the sequence implied in the Tell X example, in which new products are launched and eventually die out. We refer to the principal stages as development, introduction, growth, shakeout, maturity and decline.

It is important to grasp the three main levels on which this works. These are illustrated using the TELL X example above.

At market need level: Is the need for information systems which can transmit and receive urgent information growing, declining or what? Strategy at this level may well be determined at headquarters level, acquisitions of new companies might be made to gain a foothold in a new market, or existing companies sold off in areas no longer of interest.

Figure 3.3 Principal stages of the product life cycle

At product level: Is the demand for telex machines growing or declining, both in real terms and as a percentage of the market for urgent information systems? Strategic questions would include: What other product

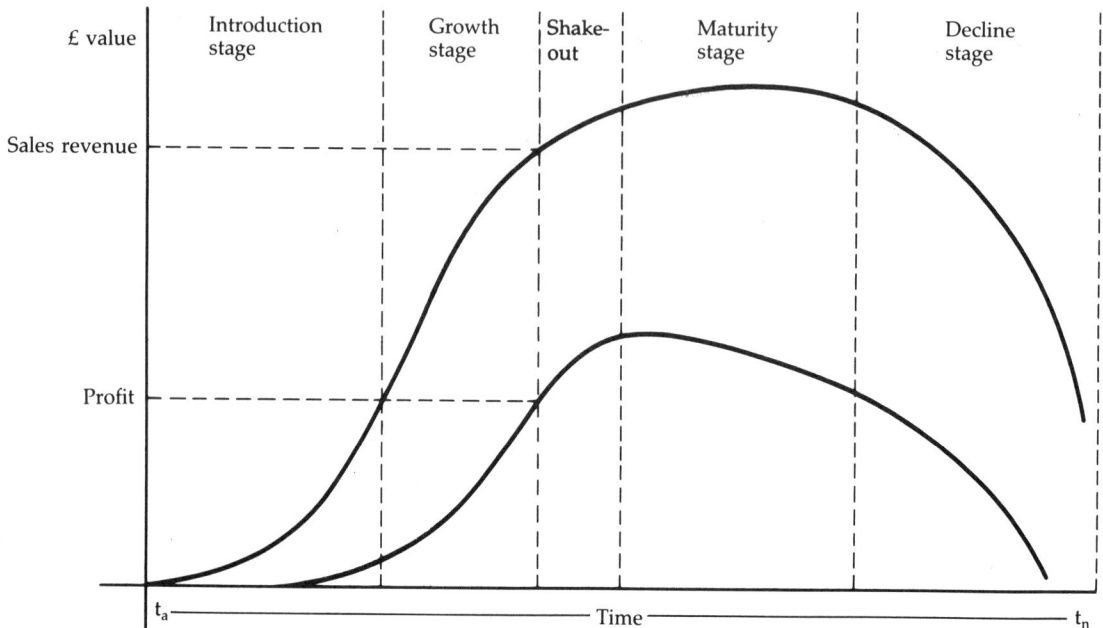

innovations, such as fax machines, might threaten telex machines? Strategic decisions here would be made at HQ or SBU level, as firms attempt to keep up with the latest trends. Failure would invoke a decision at HQ to wind up or divest the SBU.

At brand level: What is happening to the range of telex machines marketed by a particular company, both in absolute sales volume and in terms of the total sales of all telex machines? How do Tell X's product offerings compare to those of their competitors? Strategic decisions here could well be marketing and engineering decisions to change elements of the marketing mix to provide value to changing market segments. Failure could result in a decision by senior management to drop or replace a particular brand.

Mistakes in these distinctions, or failing to observe these distinctions at all, can have disastrous consequences. Classically, the failure of the Hollywood movie companies during the 1950s is often explained by the idea that the companies were preoccupied by the movie business, competing with each other, rather than focusing on the leisure industry, competing with radio and television media. Similarly, the recent failure of Wang may have much to do with their perception of themselves as makers of quality word processors, competing against other word-processor manufacturers, rather than in the information manipulation business, competing against all producers of information technology.

In this context, lateral thinking and creativity are needed to scan and understand the widest possible view of markets, scenarios, threats and opportunities.

The demand for information transmission systems grew throughout the period illustrated by Figure 3.4, but demand for telex machines went through its full cycle from introduction, when it was the smart new

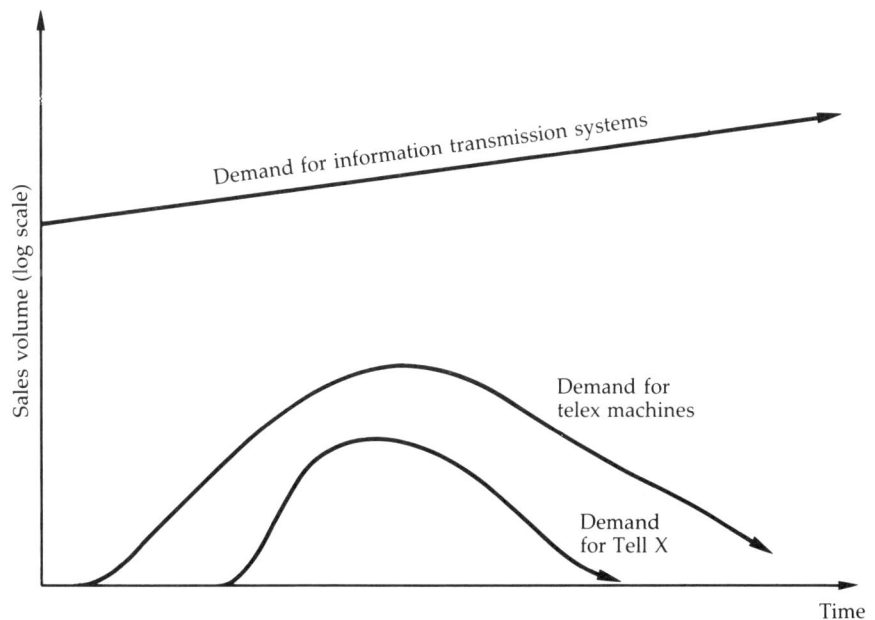

Figure 3.4 Telex and Tell X machines within the product life cycle

machine to replace the telegram, to decline, when the Fax became established as a superior product in for most uses. The figure shows that Tell X was a relatively late entrant to the market, but did well since the growth of sales for Tell X was faster than that of the market as a whole; implying that market share was growing. But sales seem to have gone into decline a little earlier than the overall market began to collapse and Tell X was, perhaps, one of the first companies to disappear.

The product life cycle is characterized by a sequence of stages, generally shown as changes in sales trends. As each change is caused by real events which should be picked up in a SWOT analysis, strategy should follow the stages in the cycle. The stages are outlined below, and identified in Figure 3.3.

First stage:
Development

In this stage, new products are conceived, designed and developed. Development costs may be relatively small, in the case of a new brand with trivial improvements on the existing products, or massive, in the case of a completely new product based on new scientific discoveries and new technologies. The engineer is commonly of paramount importance, since it is at this stage that new products are devised and developed. It is also at this stage that the marketing concept must be remembered, each new product must create more value than its predecessor, rather than simply offer more features or extras of dubious value added. This is done by extensive screening of all new ideas, to assess what the commercial benefits of the new product might be, and the ability of the organization to being the idea to a successful product launch. Generating good ideas is rarely a problem; identifying ideas which would lead to a profitable return is much more difficult.

Second stage:
Introduction

At this stage, the new product (or brand) is launched onto the market. The actual launch will be normally preceded by product testing and prototyping, as well as market testing, in which advertising literature, pricing policy etc. will have been tried out, possibly in a test launch on a small scale.

It is wrong to assume generally that products are immediately successful. Very many new products, in practice, are unsuccessful, and withdrawn from the market. But even ultimately successful products are often slow to gain acceptance. Some studies suggest the existence of an adoption/diffusion cycle, in which different types of buyer experiment with, and eventually adopt, the product.

Assuming the company has some notion of market size, it should be possible to estimate the numbers of innovators and early adopters who will first try the product, thereby influencing the early and late majorities. Only when these majority groups begin to buy the product will sales accelerate. Much can be done to influence the innovators and early adopters by the promotions element of the marketing mix, and, in the early stages at least, promotions strategy might be aimed at these groups.

Third stage:
Growth

Once a new product or brand has been adopted, there is a tendency for distribution chains to begin stocking the product in greater quantities. This rising demand can place pressure on production departments and lead to

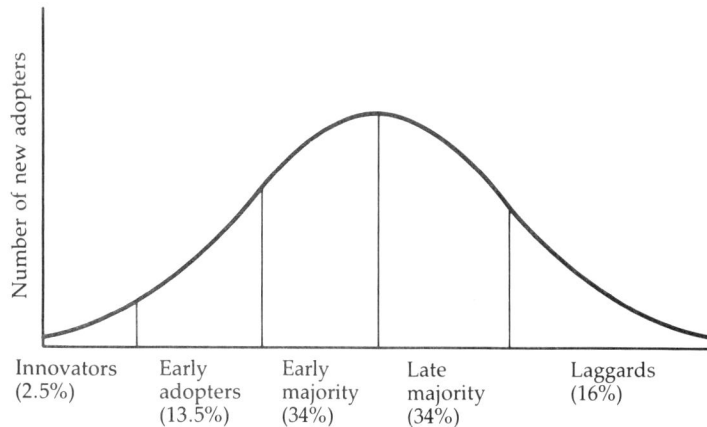

Figure 3.5 The adoption/diffusion cycle

Innovators (2.5%)   Early adopters (13.5%)   Early majority (34%)   Late majority (34%)   Laggards (16%)

shortages in distribution stocks. However, it can be a great mistake to increase capacity rapidly, as this growth in the channel chain may not reflect sales to the consumer. In some respects, Polaroid's launch of instant movie film and camera (a competitor to video cameras) suffered from this, leading to huge over-capacity problems when the product quickly flopped. Modern production methods, such as JIT, and tougher inventory control (see Chapter 5) can do much to reduce these risks.

The growth stage is extremely important. The firms showing early dominance, in terms of total sales and share of the market, will gain favourable access to distribution chains, generate early brand loyalty and benefit from any large-scale production advantages quickly (see Chapter 4). Establishing early dominance over competitors will pay dividends in later stages of the product life cycle. A rough guide to the advantage built by a company can be inferred from market share relative to the market leader. Thus, a relative market share of 0.5 implies that market share is half that of the market leader, and therefore competitive advantages are accruing to a competitor at a faster rate. Similarly, a relative market share of 2 implies that a company is market leader, and twice as big as its next largest competitor. Opportunities to gain major strategic advantages are present. Relative market share is closely monitored for this reason, although the major difficulty of identifying markets and competitors will remain.

There are a wide variety of tactics for developing dominance, including low pricing in the final market, generous discounts to the channel chain, high expenditure on promotions and continual product modifications to create additional market segments and niches. A strategy which used such tactics is commonly called a rapid penetration strategy. Such a strategy leads to expenditure on pushing the brand in excess of the revenue generated by sales, so an organization cannot use this strategy in the long term, unless there are other profitable products to cross subsidize.

Gaining a dominant market share is thus a major strategic objective, even though it may not lead directly to profits in the short term. An alternative indicator often quoted as evidence of success is the growth of sales revenue. However, this can be misleading, since it may fluctuate wildly. Moreover, when growth of sales revenue is less than market growth, market share must be falling; usually a warning of trouble ahead.

Naturally, some brands will fail to establish a significant market share. In many circumstances, this could be predicted by a SWOT analysis. For example, a company might be pleased by high sales of a particular new brand or product, but a SWOT analysis could reveal that the brand is technically inferior to that of its competitors. Further research might show that these sales are due to buyer ignorance about the relative strengths of competing goods, and general supply shortages. Assuming the company could not close the technology gap on its rivals, there are only two choices. First, the company could opt for a skimming strategy, whereby prices are raised and development and promotional expenses are cut, to maximize profits on each sale before the market becomes more competitive. Hopefully, these profits will be sufficient to develop expertise in other products and markets. Alternatively, skilful manipulation of the marketing mix may enable technically superior products to be defeated, by for example, a better product offering, a price war or using muscle in the distribution chain. Such tactics have seen the technically inferior VHS videotape displace the better Betamax product in the home video market.

Fourth stage: Shakeout

At this stage, the market growth begins to slow as channel chains are completely stocked up and new market segments are no longer large enough to develop and exploit. Price wars may break out. The weaker products; those with limited distribution, high production costs or poor design or production are often eliminated. Surprisingly, it is common for the original innovators and inventors to disappear at this stage.

During the mid-1980s, the growth in sales of home computers saw a rash of small British companies starting up; Acorn, Dragon and Oric to name a few. All had disappeared within a few years, or had been rescued from certain disaster by larger companies. In high technology products, and many others besides, the period between introduction and shakeout has become ever shorter, leading to greater pressures on innovators and inventors. But it is also worth pointing out of the three failures mentioned above, only Dragon were producing a technically weak machine. Oric's Atmos computer, although technically superior to many of the machines which survived shakeout, failed through poor strategic decisions. Acorn, again with technically good products, suffered from a wide range of production and marketing problems. Similarly, the marketing techniques applied to VHS videos enabled this technology to replace the earlier, and technically superior Betamax machines. The PC market may well be in the shakeout stage now (see Case Study 2.1).

Survival in shakeout largely depends on either a successful penetration strategy in the growth phase of the market (based on cost leadership or differentiation) or dominance in a particular market niche (a successful focus strategy). Once it is clear that a company will lose out, there is often little that can be done, other than collaboration with another, more successful company.

Maturity

This phase is characterized by slow, or no growth in total sales in the market. Any attempt to increase market share must be at the expense of a competitor. There will be a small number of large firms, accounting for most of the market sales, but a number of smaller companies may survive in local markets, or where they have identified a niche market (or small segment) not targeted by a larger rival.

In this stage, major product innovation and advertising support may be substantially reduced, and large profits made. Products and brands will generate profits to support the development of new products in the introduction and growth phases of the product life cycle. However, a close watch must be kept on market share and particularly share relative to the main rival. Should either of these begin to decline, the company must be prepared to go on the offensive with sales promotions, product relaunches and, if absolutely necessary, price cuts, otherwise the brand will go into premature decline. This involves monitoring and at least maintaining (but preferably improving), market share and profitability, and the marketing department is certainly prime in this area, as seen in Chapter 2. But the preservation of the organization's strategic health, is the ultimate responsibility of strategic managers. For instance, in a recession, the alarm should go off if revenues are dropping faster than the rate at which the market is slowing down, indicating that share is being lost, and hence another organization(s) is encouraging customers to switch.

At this stage, engineers may be involved in relatively small-scale design changes or reformulations. Often, these changes are important in their own right, and extend the life of the product. Such product extension can be commonly seen in the automobile market, when styling is changed to fit with current tastes and fashions (contrast the shape of the original Ford Escort with the recently relaunched model) or some other change induced by safety or environmental legislation.

Other redesigns might be significantly more cosmetic; aimed at giving marketers something to say in their advertising message.

Decline

Total sales in the market will fall, as consumers find ways of satisfying needs and wants in other ways. Products and brands may actually disappear. For example, the pocket calculator has displaced the slide rule at the product level. At the brand level, as demand in the whole market declines, some brands will come under increasing pressure as production volumes and prices fall.

In such cases, there are few options open. A firm may decide that there will be a large residual demand after its competitors have dropped out, and play an end game, hoping to hasten its competitors' departures. This would involve cutting prices and increasing promotional support. Other producers may opt for a harvesting strategy, in which all unnecessary costs are taken out and selling continues on its own momentum until the product is no longer profitable to produce. Other producers may decide to leave the market by selling off (also called divesting) their SBU or brand.

It is worth pointing out that some spectacular recoveries have been made from a decline position. For example, motor-cycles showed all the symptoms of decline during the 1950s and 1960s, but Japanese products galvanized the market and restored a growth position, resegmenting the market by targeting different customer groups, and making the technology affordable for new markets.

Strategy and the product life cycle

As products move from one stage of the cycle to another, marketing strategy, and the use of the marketing mix should adjust to follow suit. Sticking to one strategy throughout is almost inevitably doomed. Strategic objectives should be aimed at building competitiveness through the whole length of the cycle, either by going for rapid volume growth to achieve cost advantages, and ultimately cost leadership, or by establishing a differentiated product early on, either in the wider market or a niche market. Promotional, manufacturing and engineering activity will tend to follow the cycle, and costs must be correspondingly controlled to maintain competitiveness regardless of the strategy adopted (see Chapter 10). This gives some advance warning of the likely level of costs necessary to support products in the future, so that corporate strategies can make some assessment of whether or not successful SBUs will be in a position to generate sufficient profits to meet these costs.

At the corporate level, the organization should aim to develop a supportive portfolio of SBUs, so that profits from the products in the mature stage (and possibly decline stage) can be used to support new product development and SBUs in the introduction and growth stages.

An organization would fail its strategic health check if it found that its present product portfolio would require more support than is available, or if long-term decline was likely to weaken a large proportion of its business activity. The effects of such a failure might not be evident for some time, but the idea is to plan for the future before it becomes a disaster.

Summary of Stage One of the strategic management process

We have looked at the idea of the first stage of the strategic planning process, the strategic audit. This stage attempts to answer the question; 'Where are we now?' The audit consists of a PEST analysis, which examines the external factors affecting the organization, and SWOT analysis which tries to identify and match opportunities and threats facing the organization to strengths and weaknesses. These strengths and weaknesses are identified by departmental (or functional audits) and highlight the distinctive competencies upon which the organization must develop competitive advantage and exploit the available opportunities. Further, the analysis should identify areas of poor performance, which would limit and undermine action based on exploiting these distinctive competencies, and would also prevent the organization from responding to threats effectively. Finally, we examined the idea of a strategic health check, and discussed the product life cycle as an example of a model sometimes applied in such analysis.

Stage 2: Where should we be?

In practice, this stage is often indistinguishable from Stage 1. Many of the strategic auditing techniques are also designed to generate strategic options and alternatives. For example, by identifying that an SBU is in the growth stage of the product life cycle, the strategic options are to penetrate or skim the market. An SBU with products in the decline stage could consider harvesting, divesting and end game options. In all cases, the marketing mix and engineering activity would be adjusted accordingly. At HQ level, the idea is to produce a clear mission statement, for each SBU, which defines the markets in which the organization believes it operates, the desired position in the market and the means by which distinctive competence might be expressed in practice.

The important idea to grasp is that the mission statement determines much of what the SBU does and, consequently, the resource base and culture of the organization. However, this will only happen if the mission statement is believable (for example, it is not over-ambitious given the resources at the organization's disposal), well communicated, in that everyone in the organization regularly sees and understands it, and consistent, in that managers at all levels base their decision-making upon it.

The second activity undertaken at this stage of the process is the setting of objectives at SBU level. These are normally quantitative in nature, often specifying some financial target such as return on capital used, some growth of sales revenue or similar. Strategic marketing objectives, such as growth of market share, must also be specified. Clearly, these financial and strategic objectives must be manifestations of the mission statement, and can be adjusted as market and financial conditions change. It is important to distinguish between the financial objectives set here, and the means by which those objectives are to be reached. These financial objectives are set by HQ (although consultation and negotiation between HQ and SBU managers is an important part of the process), partly as performance targets to assess the efficiency and effectiveness of the SBU, and partly to co-ordinate and control all the SBUs under the control of HQ.

The means by which these objectives must be reached may be determined by senior managers at the SBU in some organizations, but may also be decreed by HQ in others. Obtaining commitment to the mission, and plans generated from it is far more difficult in practice than implied here, particularly if radical change is involved. The idea of change management, the problems it creates and some of the means by which these may be eased are discussed in the final chapter. As well as internal resistance to the mission, it is certain that competitors will resist intrusion into their existing markets, and may well have aggressive missions and objectives of their own.

Missions and competition

The importance of competitiveness in achieving ambitious missions cannot be overstated. The example, shown in Case Study 3.4, should serve to illustrate the point.

Case study 3.4: Competitive strategies

Car wars

There are four companies in a local market for automotive parts. Each has a different mission, and this determines much of the style and distinctiveness of each company.

Company Alpha might define its mission as becoming the largest company in high quality customized components, priding itself on customer service and attention to detail. A second, Company Beta, might aim to become the major player by producing the cheapest components, with no attempt to compete with Company Alpha in terms of customer service. A rival, Company Gamma, might set its mission to become the major producer of mass produced, low technology components, but with extra attention to after-sales and customer service. A fourth company, Delta, might be much smaller than the others (who are of roughly equal size) and produce only heavy duty parts for a local producer of rough terrain vehicles; with a modest mission statement to match.

At first glance, it would seem that the four companies have developed strategies which would prevent outright competition from threatening any one of them. Company Alpha has gone for a differentiation strategy. The products it makes, and the specialist services it offers would be completely different to those offered by the others. Company Beta has gone for a cost leadership strategy. By only producing products which can be standardized, mass produced and require little after-sales service, it can keep down costs and either make large profits, or provoke a price war to see off a competitor. Company Gamma has selected a half-way house, and would make components which can be largely standardized, but may need pre-sales or after-sales service. Company Delta has gone for a focus strategy based on differentiation, by producing only components which need to be particularly robust.

The four firms would operate quite differently in most major respects. Company Alpha might well be engineering led, employing a large number of skilled engineers and designers in both pre-sales service, application advice, production and after-sales service. Marketing activity is likely to be direct, since personal contact between sales engineers and customers is required by the nature of the product. The company may also invest heavily in specialist plant and equipment to extend the range of its services. Company Beta is more likely to be a 'screwdriver' or 'metal bashing' affair, using the lowest acceptable level of skill possible and the most cost efficient production techniques. Here, the accountant is likely to be the leader in the drive to reduce and control costs. Stocks will be kept as low as possible, to reduce the amount of cash tied up in the production process at any one time. The firm may well sell through intermediaries, such as industrial wholesalers, and have little direct contact with customers.

Company Gamma may show some elements of both, needing both a low cost production system and some highly skilled engineers for its more specialist areas.

Company Delta is likely to resemble company Alpha, although on a much smaller scale.

In the short term, it is likely that there will be only limited competition between the four companies, and the situation may well be stable. But if the market growth begins to slow, the companies may be forced to compete in order to achieve growth. Of course, there is no reason why missions should not be aimed at maintaining or defending a market position, rather than growth, but in the medium and long term there seems little point in having a mission if it is not going to stretch the organization in some way. If competitive rivalry were to increase in this market, Company Gamma would look most vulnerable, as its market segment is the least well defined, and both Company Alpha and Company Beta might target part of its business. Company Alpha would have an competitive advantage in customization, and Company Beta would have a similar advantage in the low cost segment. Similarly, Company Gamma may target either of the other main players by specializing further, particularly if either has drifted from its distinctive competence.

In the longer term, several scenarios may develop. For example, if the market for low cost, standardized components is still growing quickly, while that for customized parts slows down, Company Beta may generate sufficient cash reserves to take over Company Gamma. The combination of a low cost base plus the addition of customization and customer service expertise from the acquisition could mean that Companies Beta and Gamma together could match Company Alpha's differentiation (thereby removing it) and destroy their rival by establishing a cost leadership in Company Alpha's own market segment. Conversely, faster growth in the customized segment may allow the same possibility to Alpha at the expense of Beta. Only company Delta seems relatively safe in its own niche.

The Car Wars strategy shows how generic strategies would work to ease competitive pressures within the market, and how these strategies might be used to attack a rival. In reality, each company would also have to consider the other competitive forces, namely: the power of the buyers and suppliers, and the threat of new entrants and substitutes.

The three approaches, differentiation, cost leadership and focus, are the three generic strategies. They set broad parameters in which the firm operates. There is considerable evidence to suggest that firms which achieve either cost leadership or successful differentiation will generally do better than organizations which do not opt for, or succeed in, establishing either. But, as the example indicates, the situation is rarely stable, and opportunities and threats evolve continually.

Summary: In Stage 2, HQ will ultimately produce mission statements to set, in general terms, the long-term strategic direction of each SBU. In addition to its mission statement, each SBU will be given specific measurable objectives to achieve; particularly regarding financial performance and growth. Successful attainment of these objectives will depend, in a major part, on competitiveness in the marketplace, based on one of the generic strategies. However, having a cost or differentiation advantage still does not guarantee success, these advantages must be deployed by a successful strategy designed to achieve the targets set. The means by which these objectives are to be reached is the main task of the third stage.

Stage 3: Generating options

Stage 3 answers the question 'How are we going to get there?' The options generated should be aimed at achieving the mission statement (Stage 2) with the resources, strengths and opportunities available to the organization (Stage 1).

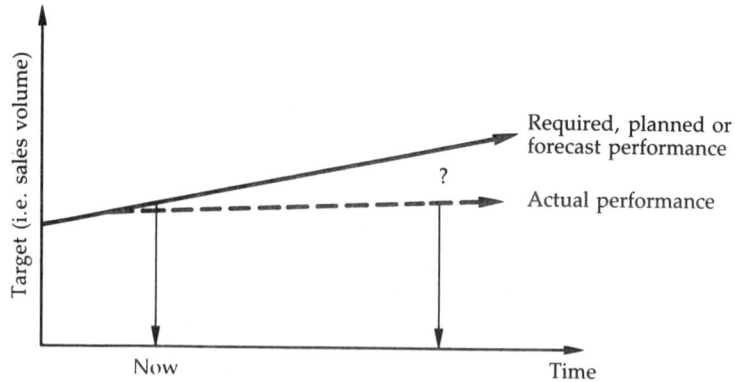

Figure 3.6 Gap analysis

So, at this stage, the major preoccupation is with generating, evaluating and selecting options by which the SBU will achieve its objectives. A good starting point for this is gap analysis, which looks at the difference between actual and predicted performance according to some performance indicator.

Having identified a gap between actual and forecast performance, the SBU has three options:

1 *Do nothing*

This option would be selected if the gap was trivial, or if action to correct a gap would lead to problems elsewhere; for example, if the cost of closing a short-term gap in sales volume targets were thought likely to create greater gaps in financial performance.

2 *Lower objective*

It would be easy to close the gap by reducing the objective set to meet actual performance. This option would be selected by an unambitious management. An acceptable alternative might be to lower the objective in response to changes in the business environment, such as an import quota imposed by a foreign government, or a sudden rise in the exchange rate which greatly reduces competitiveness in a traditional market. However, such adjustments in objectives should always leave a gap for SBU management to close.

3 *Close gap*

This option requires management to change its current business strategy and operations in order to grow faster and achieve more than would have been the case without change.

In all but exceptional cases, the purpose of this stage is to devise a means by which this gap can be closed. To a large extent, the search for such means can be simplified by use of the product/market matrix; also called the Ansoff matrix after its originator, who argued that growth opportunities came from combinations of current market (customers) and new markets, current products and new products (see Figure 3.7).

1 Market penetration: old customers and old products

A market penetration strategy would involve wrestling market share away from a competitor, without modifying the product in any way. This could be done in several ways, including increasing the usage rate of existing customers and encouraging non-users to try the product. These techniques

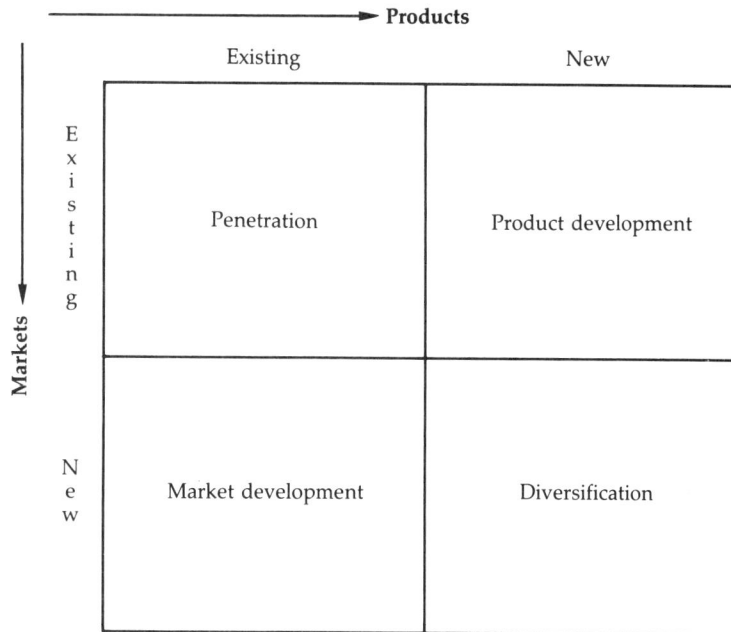

Figure 3.7 The Ansoff matrix

have the advantage of increasing share without directly facing a competitor, and are widely used in the growth stage of the product life cycle. A third method would involve picking out a weak competitor and positioning the product in direct competition to it, although this clearly carries significantly greater risks. In a mature market, or where there may not be any significantly weaker competitors, market penetration opportunities may not be sufficiently large to close the gap, and so other more subtle marketing strategies must be considered.

2 Market development: old products and new customers

A market development strategy is one based on selling existing products into new markets. These new markets may be different segments, such as industrial rather than consumer markets, quality rather than mass markets, young buyers rather than old, and so on. Alternatively, the new markets may be physically remote, such as new regions or countries to enter and export to. Third, the marketing mix may be developed to reach new markets, perhaps by using different distribution chains to reach new consumers, or by offering products on lease terms. This strategy is attractive since it requires little or no product development; often a change of promotion strategy is sufficient to entice trials from a new market segment. However, as with market penetration, there may be limited opportunities for this technique in mature markets.

3 Product development: old customers and new

Product development requires the organization to sell new products to its existing client base. There is much advantage to be gained by selling products in a segment where distribution chains already exist and some degree of brand loyalty may have been won already. In many cases, the company will attempt to close the gap by designing, developing, producing, distributing

61

and promoting the new products itself. In other cases it is possible to factor the new products, that is, to sell products completely or largely made by suppliers under a brand name familiar to consumers, albeit for other products. In practice, a good distribution chain into, say supermarkets, may be more advantageous to a company launching a new product than years of production and development experience.

The maintenance and support services of the original product offering may also be developed further, so that such services are offered on products not actually made by the company. On the other hand, brand loyalty does not always transfer well, and disappointing new products may actually undo some of the good marketing of the original product.

4 Diversification: new products and new markets

The fourth strategy is that of diversification, that is, where neither products or markets are familiar to the company. Such a strategy is fraught with risk, requiring both technological and marketing innovation. Possibilities would include activity in related markets, such as taking over a distributor or a supplier. Such activity may well be part of a competitive strategy, as well as closing a performance gap. A second possibility might include exploiting related technologies, that is, applying the company's technological expertise to new products. Third, the company could move into areas where it has neither related marketing experience or technological expertise.

Summary: In the third stage the SBU will outline general strategies designed to close the gap between objectives set and predicted performance. The plans will generally be based on exploiting competitiveness based on a successful generic strategy. It now remains only to generate specific operational plans to put the strategy into effect, which is the main task of the fourth stage.

Stage 4: Operational plans

This stage attempts to answer the question, 'What are we going to do now?' At this stage, departments will actually have to consider programmes, take decisions and commit resources to carrying out one or more of the strategies adopted in Stage 3. Senior management, after due consultation with all departments, will produce a series of departmental objectives. Individual departments will then carry out these plans.

Since these plans will actually be acted upon, much more is required than the general financial and strategic targets generated in Stage 2. The acronym SMART is sometimes used to represent the necessary components of such operational objectives, standing for:

S: *Specific*: The activity will be clearly and accurately defined – in the fewest possible words – and individuals responsible for carrying out the task are clearly identified.

M: *Measurable*: The objective must generate activity which is measurable in quantitative terms; such as the number of sales visits, number of new designers appointed, financial savings on debt management etc.

A: *Achievable*: There is no point in setting targets and goals that clearly are not within the bounds of possibility in the short to medium term. Considerable managerial skill is needed to judge between an objective which stretches, and an objective which is ignored because it is unrealistic or even absurd.

R: *Resourced*: What is sensibly achievable will depend on many issues such as: level of training possible in given timeframes, learning curves, and financial investment available for essential support functions, to name a few. A prime function of management is to negotiate with people who are going to do the job, and to set budgets which enable the activity to be carried out effectively (see budgeting in Chapter 8).

T: *Timeframes*: Actions planned to a deadline, or series of deadlines depending on complexity. Hence, these are negotiations to establish just what rates of achievements are possible, in what timeframes results are to be monitored, and what remedial action is to be taken to close any unforeseen gaps. The person concerned is subsequently clearly accountable for making the deadline. This will apply to all levels of function, from the chairperson downwards.

Departments will thus need to draw up action plans to facilitate and co-ordinate all the tasks necessary to achieve these strategic objectives. By way of example, marketing department activity is likely to consist of managing the marketing mix (or four Ps, see Chapter 2), and undertaking marketing research toward achieving departmental objectives. Each action or aspect of the plan should be written in SMART terms.

So, for example, Tell X might have been instructed to withdraw their telex machines from the UK markets and develop the South American market (an example of market development). Senior management might assess methods of achieving this, and elect to use foreign distributors as part of the marketing mix. The marketing department would have had a major input into the strategic decision, and bear much of the responsibility for making it work in practice. Part of the operation to do this could be covered by the following objective:

Ms Hays, the marketing manager, will produce a short list of ten suitable distributors for the Peruvian market within four weeks with a budget of £XXX.

Clearly, Ms Hays will have to generate a larger list of distributors from which to make her selection. This task may be delegated to an assistant, working to complete the objective:

Ms Hope, assistant marketing manager, will identify forty distributors with coverage of all major business centres in Peru, within three weeks at a cost of £XXX.

Ms Hope may actually do the work herself, delegate part of it to a junior, or commission an external research agency or consultancy to carry out the secondary research necessary. In any event, the instructions given will be characteristically SMART.

Every department will be affected by Tell X's new strategy. The marketing department will have to commission marketing research to establish segmentation and position, as well as identify, interview, assess, appoint and motivate new distributors. Moreover, promotional literature, sales conferences, pricing and strategy will require development and execution. The engineering department may well have to adapt the product to suit differences in local telecommunications systems, climate, language, electricity supply and safety laws. The production department will need to adapt current manufacturing processes to incorporate these changes. The personnel department may be involved in the recruitment and training and language tuition of personnel offering support and maintenance to the Peruvian distributors, and the redundancy implication of the decision to wind down the UK sales operation. Accountants will draw up budgets to cover the costs of these activities. All activities within every department can be directed by SMART objectives set out in the way outlined above, hence the expression, 'management by objectives' (MBO). MBO was the most popular method, during the 1980s at least, for allowing mission and strategic objectives to cascade down to the most humble operations activities, linking them together in a way which may be completely invisible to the operative concerned, but none the less unifying the reinforcing overall strategy and commitment to it.

In practice, things rarely go as smoothly as implied here. Short-term operational crises often blow MBO off course, as will failure to obtain commitment to the strategy before operational plans are made. Interdepartmental politics and misunderstanding may also generate objectives which protect vested interests, rather than push through the new strategy. There is much that can be done to encourage such commitment, for example, setting objectives by negotiation rather than imposition, and linking incentives to strategic rather than operational activity. Multi-disciplinary teams may be set up to bear the brunt of planning and operations, so as to lower inter-departmental strife and jockeying for position. However, the principal barrier would generally appear to be the corporate culture, rather than technical or communications problems; important as these are (see Chapter 12).

Summary: In Stage 4 things actually get done, as management generate a series of SMART objectives, which staff can use as a basis for action. Although much can go wrong, management by these objectives offers the possibility of achieving the strategic targets set in earlier parts of the planning process.

Stage 5: Monitoring and evaluation

The final stage answers the question 'How will we know when we've got there?' In reality, it's more likely to indicate the point at which things have begun to drift off course; and as such is crucial.

If measurable objectives have been set throughout, this question might appear to be straightforward; a successful strategy is one which reaches the objectives set. Continuous monitoring and evaluation must take place to ensure that the plan, and its constituent activity, remain the basis for action plans and resource allocation.

For example, in the Tell X example, suppose three distributors had been appointed to sell the products in Peru. Distributor A was appointed to sell

Table 3.1 *Volume of Tell X Telex machines Sold in Year 1*

	Target sale per quarter	Actual sales quarter	Variance
Distributor A	3000	3120	+4%
Distributor B	4000	6450	+61%
Distributor C	5000	1830	−63%
Total	12 000	11 400	−5%

to government offices and institutions, Distributor B was selected for its existing contacts in seaports and coastal regions, while Distributor C was expected to open markets in the interior. After one year, the overall sales target was just 5 per cent lower than the ambitious target set when the entry plan was first conceived.

The marketing department may have a range of possibilities to close this gap, as outlined above, but overall, Tell X may believe itself to be close to achieving its objectives. However, small corrective actions and a smug attitude may be premature if the company has not identified the major cause of the gap. Analysis will be greatly facilitated if measurable objectives have been set throughout. The variance analysis above may be overlooked if the gap is small, or if targets have actually been reached.

When examined in this way, a small performance gap is only an indicator of a much larger problem, concealed by the fact that the variances largely compensate for each other. Further analysis would be required to examine the reasons for these variances, and, in particular, whether Distributor B's over-achievement is in any way connected with Distributor C's equivalent under-achievement. It may be that Tell X have greatly misjudged the size of each segment, or the competitive pressures faced. Some elements of the marketing mix in the interior may be fundamentally wrong. Alternatively, Distributor B may have begun attacking the interior, at the expense of Distributor C. Whatever the reason, the broad strategic options generated by gap analysis and the Ansoff Matrix are unlikely to be helpful at this stage; the problem is clearly a tactical one requiring a review of the marketing mix used to achieve the strategic targets.

At a higher level, it is possible for all the operational targets to be met, but the SBU to fall short of its target return on capital, or some other target set by HQ. It is in this instance that a return to gap analysis and review of strategic options is appropriate; particularly if a strategic audit reveals that the external environment has changed in some way that affects the realism or desirability of the original strategic objectives.

The monitoring of operational and strategic objectives should be routine and continuous. Variances should be investigated, even when they appear small, such as in the example above. Analysis of financial variances is examined in much greater detail in Chapter 9.

Summarizing the steps

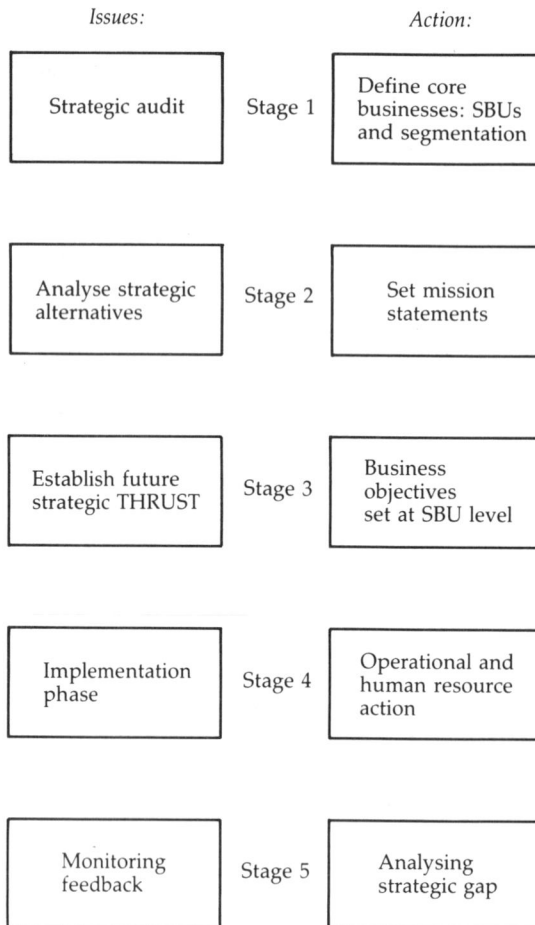

Issues: *Action:*

Strategic audit	Stage 1	Define core businesses: SBUs and segmentation
Analyse strategic alternatives	Stage 2	Set mission statements
Establish future strategic THRUST	Stage 3	Business objectives set at SBU level
Implementation phase	Stage 4	Operational and human resource action
Monitoring feedback	Stage 5	Analysing strategic gap

Figure 3.8 The strategic planning process

The strategy process

The process of answering the five strategic questions should be considered a normal business process, rather than occasional or exceptional. As the business environment changes ever more quickly, the planning process needs to respond by resetting objectives and strategic processes more quickly and urgently, using for example real time IT systems.

Increasingly, the emphasis in a modern business is changing from a ponderous planning and budgeting exercise, to strategy process for making and implementing decisions very quickly. Organizational structures, hierarchies and planning processes which prevent rapid response are now becoming obsolete at a frightening pace.

Chapter summary

Organizations should monitor the environment in which they operate, and assess the opportunities and threats observed in terms of the strengths and weaknesses of the firm itself. Strategies should be devised which fit the organization's strengths to those opportunities, while minimizing or eliminating the effects of weakness where possible. Strategies can be put into operation by use of SMART objectives, and the outcomes of plan imple-mentations should be monitored against operational and strategic objectives.

Further reading

There are many books on the business environment, all of which date very quickly as the environment changes. Many traditional books on the subject appear to be a mixture of law and economic theory, and are of limited practical help. One very valuable exception is *The Business and Marketing Environment*, by Palmer and Worthington (McGraw-Hill, 1992) which tries to draw marketing implications from its general descriptions, and relate these to business objectives. To keep up-to-date however, there is no alter-native to studying such quality newspapers as the *Financial Times*, and listening to informed journalistic opinion in all media. Economic experts and their forecasts are also worthy of attention, as they are always well argued, based on good information and sometimes accurate. Such a background knowledge would make J. Dudley's book *1992, Strategies for the Single Market* (Kogan Page, 1990) and *European Business Strategies* by R. Lynch (Kogan Page, 1990) extremely useful for adding an international dimension.

There are many accounts of the strategic planning process, and the role of marketing within it, including *Marketing Management, Analysis, Planning, Implementation and Control*, by P. Kotler (Seventh Edition, Prentice Hall, 1991). *Marketing Plans*, by M. McDonald (Butterworth-Hienemann, 1989), and *Market Led Strategic Change*, by N. Piercy (Butterworth-Hienemann, 1992) are more coherent, practical and up-to-date, while the authors also retain an affection for *Offensive Marketing*, by H. Davidson (Penguin, 1987).

The definitive works on the competitive environment, generic strategies and building competitive advantage by studying value creation are Michael Porter's books *Competitive Strategy* and *Competitive Advantage* (published by Free Press, 1980 and 1985 respectively) which are quoted extensively throughout the literature on business strategy. However, these books are heavy reading, and *The Essence of Strategic Management*, by D. Bowman (Prentice Hall, 1990) can be regarded as a more accessible introduction. Experienced technologists are referred to *Business for Engineers*, by B. Twiss (Peter Peregrinus, 1988) which develops the role of technology and technical innovation within the strategic process, while business orientated final-year and postgraduate students are referred to such standard works as *Exploring Corporate Strategy*, by Johnson and Scholes (Prentice Hall, 1989), *Strategic Management, an Integrative Approach*, by Hax and Majluf (Prentice Hall, 1984), *Business Policy*, by Luffman, Sanderson, Lea and Kenny (Blackwell, 1991) and *Strategic Management*, by Bowman and Ashe (Macmillan, 1987). An extensive discussion of various strategy models and their uses can be found in *Strategic Management, A Methodological Approach*, by Rowe, Mason, Dickel and Snyder (Addison Wesley, 2nd Edition, 1990). *Strategic Decision-Making*, by Gore, Marshal and Richardson (Cassell, 1992) is particularly good for raising doubts and expressing criticism of the various theories.

Tutorial exercises

3.1 Xylogics is a very successful producer of one product range in a fast growing market. Presumably, at some (hopefully) far distant time, the market will slow down, and may decline. What will happen to Xylogics? Is there anything that can be done in the meantime?

Case study 3.5: Funfurs plc

Funfurs plc is a large manufacturer of 'plushfibre' which is used in the production of soft and cuddly toys. It is designed to make the artificial skin or fur softer and also possesses fire retardant qualities. There are several types of plushfibre, but Funfur produces only the following four types at present:

1 Plushfibre 1, known as Velviplush; is used in traditional teddy bears, and was introduced in its original form a long time ago. It remains the mainstay of the company.

2 Plushfibre 2, known as Splush; is used in making toys which are likely to get wet. It started well when it was introduced on cuddly seals, which were sold as part of a conservation promotion. Recently, the market expanded when turtles became popular. Subsequently a producer from the Pacific Rim entered this expanding market and has launched a low cost product with improved performance.

3 Plushfibre 3, known as Superplush, is a relatively new product used in high tech and advanced cuddly toys.

4 Plushfibre 4, known as Plushette, was one of the company's original products and was mass produced for many years. It is now used on gonks and other cheap toys given as free gifts or fairground prizes.

Key statistics – from Profit and Loss and Balance Sheet

| | 1991 | | 1992 | | 1993 | |
	Turnover %	Gross profits %	Turnover %	*Gross profits %	Turnover %	Gross profits %
Plushfibre 1	48	72	38	60	34	62
Plushfibre 2	10	5	7	2	6	(2)
Plushfibre 3	24	18	45	33	55	37
Plushfibre 4	18	5	10	5	5	3
	100	100	100	100	100	100

	1991	1992	1993
Total turnover	250k	3500k	3600k
Total gross profit	1000k	1500k	1350k

Gross profit percentages are proportion of total gross profit obtained from this product.

Plushfibre I	*1991*	*1992*	*1993*
Market share (%)	24	30	40
Relative share (leading)	4x	5x	6x

Market projection: Steady decline of around 5 per cent over next 5 years.

Plushfibre 2			
Market share (%)	8	6	4
Relative share (following)	0.66	0.5	0.35

Market projection: Continued growth for next five years at 15 per cent per annum.

Plushfibre 3			
Market share (%)	60	50	40
Relative share (leading)	3x	1.6x	1.5x

Market projection: Continued growth for next eight years at 20 per cent per annum.

Plushfibre 4			
Market share (%)	7.5	7	6
Relative share (following)	0.33	0.2	0.16

Market projection: No significant growth over next five years.

Note: Relative market share gives an indication of share relative to competitors. For example, Plushfibre 3 had a share three times bigger than its nearest rival in 1991, while Plushfibre 4 had a share just one-third of the market leader in the same year.

3.2 (a) Identify the position of each product on the product life cycle. What are the future prospects for each product?

(b) What would you do with each product? Produce an action plan (based around the five stages outlined in this chapter) for one of the products.

(c) What are the financial implications of your action plan?

3.3 (a) How would you go about assessing the ethical considerations of the company's past and present activities?

(b) Is it possible to argue that the company's rescue plan is highly ethical, particularly with regard to the endangered species concerned?

(c) One of the products listed under 'cultivated flora' is hemp, a product which has both legal and illegal uses. What ethical issues are raised by this?

(d) Taking one item from parts 1 to 3 of the rescue plan, show how you would use the marketing mix to market the product.

(e) Write an action plan for launching one of the company's new 'ethical' products.

Case study 3.6: The Colonial Foraging Company (established 1749)

The company was set up by the Coltrane family as British trading influence and empire expanded throughout the world, to provide a foraging service for troops and expatriates. It developed business operations throughout the British Empire, and places where British influence was strong. In the mid-nineteenth century, the company expanded its activities to include more general trading activities; in particular, it imported animal products from British colonies to Great Britain. With the reorganization of the British Army at the end of the nineteenth century, the company was forced to depend on these trading activities entirely. It responded in an innovative way, becoming one of the first major beef producers to invest heavily in refrigeration techniques. It was therefore able to supply beef throughout the empire. Links with Japan were made in the early years of this century, when the company provided foodstuffs to Japan during its war with Russia. These links were strained during the early 1940s, but were quickly re-established during the rebuilding of the Japanese economy during the following two decades.

The company survived the transition from empire to commonwealth without major disturbance, and by 1975, its product ranges consisted of the following:

	% of revenue	% of profit
Big cat fur for clothing industry	5	10
Ivory for jewellery market	8	12
Other elephant products	2	1
Whalemeat for human consumption	6	12
Other whale products	8	12
Rhino horns	1	4
Live tortoises for domestic pets	3	6
Exotic birds for domestic pets	2	3
Big cat carcasses for trophies	4	8
Wild flora products	4	9
Cultivated flora products	7	13
All bovine products	50	10

Since the late 1970s there has been increasing pressure on the company to reassess its business activities, indeed, many have now become illegal. Moreover, pressure exerted by several highly effective pressure groups has been applied to all the company's products, not merely the controversial ones. In particular, pressure on the beef market, added to overcapacity in the market generally, and the fact that Colonial Foraging's beef tends to be of rather inferior quality, has left the company with extensive surplus ranch and refrigeration capacity in several parts of the world.

The combined pressure on all of the company's major trading activities now threatens its future existence. Unless a new direction can be found immediately, the company will be broken up and its extensive overseas assets will be liquidated and its financial reserves returned to the small number of Coltrane family shareholders.

Senior management have devised the following rescue plan:

1 Surplus ranching capacity could be used to 'farm' elephants, big cats etc. for their commercial value.
2 Surplus freezing capacity could be utilized by offering a service freezing bodies of people who hope to be revived at some future date.
3 The company are interested in the possibility of developing whale farms in the Pacific Ocean.
4 The company are interested in developing a range of ethical products, in order to reverse their current image profitably. They are excited by the idea of developing an autogiro, which could be used against ivory poachers in African national parks.

3.4 What stage of the product life cycle has been reached by Borland in Case Study 2.1 on page 26. What strategies are involved, and what alternatives might have been tried?

3.5 Consider the view of the business environment taken in the *Guardian* in Case Study 3.1 on page 43. Contrast the engineering environment now, with that faced in 1992.

3.6 (a) What do you see as being the key elements of the rescue plan outlined in the General Motors case study below?
(b) Generate a strategic audit and action plan, using the information in the article and any other available to you.

3.7 Using the Case study 'Car Wars' on page 57, show how generic strategies would help each company in responding effectively to:
(a) A powerful supplier
(b) A powerful buyer
(c) A powerful new firm entering the market
(d) A development in plastics technology which allowed components to be made more cheaply

Case study 3.7: General Motors: The embattled giant

Serious surgery needed to staunch flow of red ink

Revolution, observed Mao Tse Tung, that art, is not a dinner party. It cannot be advanced softly, gradually, carefully, considerately.

Mr Robert Stempel, Chairman of General Motors, learned this to his cost this week as pressure from fellow directors forced him to resign after only twenty-seven months at the helm of the world's largest industrial company.

The tall, booming-voiced but kindly Mr Stempel stood accused of failing to move quickly and toughly enough to implement a ten-month-old revolution designed to save the company from financial crises. He agonized over plant closures. 'The message', said one analyst, 'is that nice guys don't win.'

However, the main message behind the most dramatic US board-room upheaval in many years is the sheer gravity of the outlook facing GM – a case study in industrial decline stemming from decades of insular, bureaucratic complacency.

GM is the world's largest industrial corporation with the biggest turnover , $123bn in 1991, the biggest workforce, 756,300 last year – and in 1991 the biggest loss ever recorded by a US corporation at $4.5 billion.

It still accounts for nearly a sixth of world vehicle production, while its turnover rivals the gross domestic product of a medium-sized industrial country such as Austria. Last year with US losses of $7 billion and a US vehicles sales volume of 4.3 million, it lost $1,631 on every vehicle sold in the US.

The group, which has only made a profit in one quarter since Mr Stempel became chairman, is bleeding red ink. It yesterday reported a third-quarter net loss of $753m, bringing its losses for the first nine months of 1992 to $971m. Nor is there any likelihood of a return to the black in the near future.

The outlook is sufficiently bleak for some analysts to be raising the spectre of GM eventually filing under Chapter 11 of the US bankruptcy code. That is hardly imminent, but it could become possible two to three years down the road if GM fails to turn round the source of all its problems – its core North American automotive business.

For the red ink here is overwhelming the profitability not only of GM's international car operations, but also of its three huge non-automotive subsidiaries – General Motors Acceptance Corporation (GMAC), its finance arm; GM Hughes Electronics, the aerospace group; and EDS, the data systems business.

Over the past two years the North American car operations have lost a staggering $12 billion, including special restructuring charges, and analysts expect it to lose up to $4bn this year.

Cash has been draining from the business at a disturbing rate and GM has been plugging the gap through additional borrowings and the issuance of some $4 billion of new equity so far this year. The group also has a large unfunded pension liability, which some analysts believe could reach $12 billion, or roughly a quarter of liabilities, by the end of this year.

GM's credit ratings remain reasonably good, but Moody's Investors Service is reviewing $70bn of its debt, and Standard & Poor's, the other big rating agency, says it might lower the company's ratings by mid-1993 unless the financial performance improves.

The most immediate danger for GM is that the agencies might lower the rating on its commercial paper – low-cost short-term money which GMAC relies on heavily. This would raise its borrowing costs significantly.

Boardroom concern over the debt ratings may have been a factor in the timing of Mr Stempel's downfall and the more aggressive revamping programme which now seems likely.

Yet GM's North American car problems are of such longstanding (and were acquiesced to by many of the non-executive board members now wielding the axe) that a turnround will not come quickly or simply.

GM was created in the early years of the century when Mr Willy Durant, an entrepreneur with a passion for motoring, merged together a group of fledgling car businesses. It was then gorged into an efficient bureaucracy in the 1930s by Mr Alfred P. Sloan, one of the giants of American industry.

But it so came to dominate the US car market, with roughly 50 per cent of all sales in the 1950s and 1960s, that complacency set in.

Dominated by finance men (ironically, Mr Stempel was the first chairman with an engineering background since the 1950s), it lost touch with the marketplace and was ill-prepared for the mushrooming of Japanese manufacturing plants in the US in the 1980s. In the last decade it became mesmerized by futuristic technologies and the drive to build 'the twenty-first century corporation'. Its investment of tens of billions of dollars obscured the failure to keep abreast of the basics of successful car-making.

GM's share of the car market plunged, from around 46 per cent at the start of the decade to around 35 per cent now, as its dull, poorly designed, look-alike cars, with a dubious quality record, lost ground to the Japanese.

However, the company failed to slim down, shielded from the financial implications of its crumbling market by booming US demand and its sheer size. So when recession hit in 1990 it found itself addled with hugely uneconomic factories and bloated bureaucracies.

Mr Stempel, after several early rounds of cutting, finally bit the bullet hard last December when he announced plans for the closure of 21 plants and the loss of 74,000 jobs – 20,000 white-collar and 54,000 blue-collar – by the mid-1990s.

But non-executives on the board, concerned that the revolution was not moving fast enough, staged a mini-coup in April: Mr Stempel was replaced as chairman of the board's executive committee by an outsider director, Mr John Smale, while the chairman's right-hand man, Mr Lloyd Reuss, the head of North America, was replaced by Mr John Smith, formerly head of international operations.

Under Mr Smith, who won his spurs turning around GM Europe in the 1980s, the pace of change has speeded up, but the management has faced increasing opposition, most notably from the United Auto Workers' union, which staged two strikes at local parts plants in the early autumn.

These were ostensibly over local grievances, but the so-called 'Apache raids' were widely seen as warning shots to GM over job cuts. The board evidently felt Mr Stempel did not respond to the union threats sufficiently robustly.

The position of chairman is expected to be filled, at least on an interim basis, by Mr Smale, a tough manager who revitalized the bureaucratic Procter & Gamble in the 1980s, while Mr Smith is likely to take on Mr Stempel's chief executive mantle. Their most crucial tasks are to:

- Close assembly and parts plants and use the remaining ones more efficiently. So far GM has only announced the names of fourteen of the twenty-one plants scheduled for closure. More are expected before the end of the year.
- Conserve cash. Directors may consider cutting the dividend at next Monday's board meeting.
- Cut its components costs. One of the most dramatic changes instituted by Mr Smith has been the creation of the post of worldwide parts purchasing chief. The aim is to use GM's scale economies to buy components from whatever suppliers around the world can combine good quality with the cheapest prices.

  Demands for price cuts of 20 per cent or more from some US suppliers have stirred up strong opposition, while GM's own vast network of parts subsidiaries fears it will lose business and jobs.
- Sell off non-essential components businesses. Mr Smith is expected to announce GM is willing to dispose of some large slices of its parts operations and reduce the company's vertical integration.
- Quickly cut the size of the workforce. The company has announced plans to speed up by a year its planned 20,000 white-collar job cuts, which should be completed by the end of 1993.

  It is also believed to be negotiating a deal with the UAW which would offer blue-collar workers special financial incentives to leave the company early. This could mean a considerable charge against earnings but could also reduce the potential conflict with the UAW over job losses.
- Negotiate a new labour pact with the UAW. The current three-year deal, due to expire a year from now, contains remarkably generous GM concessions: any worker it lays off continues to receive virtual full pay out of a special $4 billion company fund.

  Given the weakness of the car market, there are doubts whether the fund will last till next autumn (which gives the union a strong incentive to support the early retirement scheme), but the UAW will doubtless want the package renewed in 1993.

GM will want to negotiate cuts in its healthcare provisions, which have long given the company the nickname 'Generous Motors'. It recently announced plan to make white-collar workers pick up a larger share of their health costs.

• Rationalize its engineering and manufacturing organizations. Last week Mr Smith announced plans to cut from six to four the number of GM car engineering and manufacturing divisions in North America, with the loss of some 10,000 white-collar jobs.

The idea is to simplify the GM product range, standardize parts not visible to the customer, and eliminate the rivalry between design, engineering and manufacturing operations.

The ultimate goal is to cut the number of platforms – the chassis to which different styles of passenger compartment are mated – from nearly twenty to just seven.

There is enough ammunition here for any number of UAW 'Apache raid' strikes over the coming months, or a full-blown union/management confrontation next year.

However, over the past few weeks union leaders appear to have been adopting a more conciliatory tone and the ousting of Mr Stempel sends a powerful message of management determination.

Even if it cuts costs successfully, GM will still face the huge challenge of maintaining its market share, which depends heavily on deeply discounted sales to fleet buyers, such as car rental companies, and conservative, ageing individuals who are gradually dying off.

It has some hits, such as the Saturn, a compact car built by a separate company which imitates the best Japanese practices, but is weak in buoyant sectors such as sports utility vehicles, mini-vans and mid-sized saloons.

The turnround of its operations in Europe (where the market is weakening) show what GM can do, yet the US is a far larger and more complicated operation, and time is very short. Mr Smith's Long March could become very gruelling battle.

Financial Times, Friday, 30 October 1992

4 Traditional production methods

The next two chapters look at manufacturing strategy, and the way it can be used to develop a lasting competitive edge. In this chapter the traditional functions of Western manufacturing are covered. In the next chapter, these traditional methods are contrasted with revolutionary flexible manufacturing systems (FMS), materials management systems such as 'Just in Time' (JIT), and the concept of 'Total Quality Management' (TQM).

Traditional production methods

Bespoke production

In most industrial and consumer markets, there is a limited demand for one-off or heavily customized products. The value of these products may be relatively small, such as made-to-measure shoes or suits, or very large, such as a ship. Here, production methods tend to rely on traditional craft skills, and are prohibitively expensive in most instances. Only where large-scale methods are impractical are such techniques first choice, and even then standardized and off-the-shelf components are used wherever possible. In other instances, a desire for quality and distinctiveness – in such products as the Morgan sports car – determines the choice of this technique.

In some respects, bespoke production could be the ultimate in marketing orientation, since each product could be designed for each individual customer.

Large scale production

Western industrialists have long held the view that large-scale production is intrinsically more cost effective than small-scale outputs. The strength of this conviction is based, in no small part, on the economist's notion of economies of scale. Briefly, it is argued that large-scale production will enable firms to buy in bulk, use advanced technology on the production line and employ more and better specialists in the factory. Outside the factory, the firm will use its size and muscle to obtain better financial deals from suppliers and financiers, operate its own transport fleet and obtain both cheaper and more extensive advertising space in the media, unavailable to a small company, at a relatively low price. In short, the idea is an extension of the much cherished view that as output per year, (or some other time period) expands, the average cost of each unit of output will fall.

Mass production

The logical conclusion of the notion of economies of scale suggests that success will result from high output volumes. Certainly, great success was obtained by industrialists using mass production techniques in the first half of this century. The most celebrated example was that of the first mass-produced motor-car, the Model T Ford. Attempts to modify the basic model were greatly resisted by the company, even to the extent of producing them all in the same colour. Indeed, it could hardly be otherwise, since

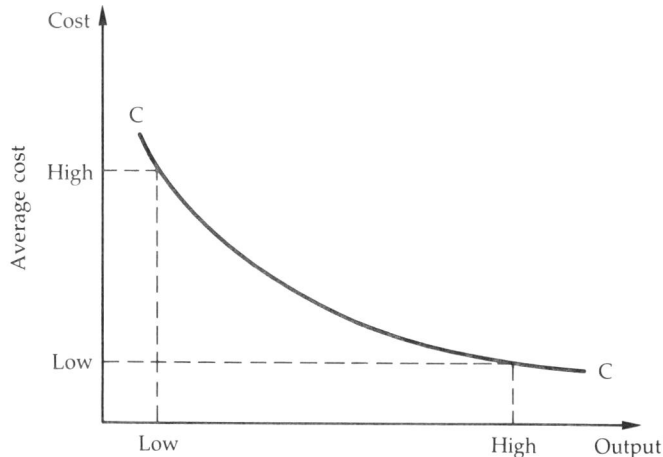

Figure 4.1 Falling average cost curve

workers and capital equipment were arranged into long, inflexible production lines, which could not be interrupted without hideous disruption and expense.

Given our discussion of market segmentation in Chapter 2, it is clear that in most markets, such uniformity of product would be unacceptable to consumers today. However, in some engineering component markets, where there is little product differentiation, mass production is entirely acceptable.

Batch production

Batch production resembles mass production in some respects in that medium to large volumes of output are created. However, in batch production, each stage of the operation is carried out separately, so that a whole batch of metal rods might be cut to length at one time, and then stored or transported directly to the grinding shop. The system is clearly more flexible than mass production, in that buffer stocks can be employed to make up any temporary shortfalls.

In some cases, where production flows continuously, it may be difficult to distinguish between mass production and batch production. The term 'continuous batch production' has been used to describe such cases, which are very common in the production of standardized products; such as bulk chemicals and sheet metal.

Traditional production systems

In reality, manufacturers use a combination of all three techniques. A final product is likely to be assembled using a flow system very much like the mass production system described above, while sub-assemblies are likely to have been made using hybrid batch production. Prototypes are generally made as one-offs, using bespoke methods. Generally these three methods are more complementary than alternative, and are used together to overcome production bottlenecks and keep costs low.

These techniques require a wide variety of planning and control techniques in order to maximize their advantages.

Planning techniques

With batch production, it is desirable to plan batch sizes that minimize costs. These costs can, broadly speaking, be broken down into storage costs and transaction costs.

Storage costs

The costs incurred by storing a batch will include:

- rent and maintenance of warehouses and stores,
- cost of maintaining accurate stock records,
- losses due to damage, deterioration, pilfering and inaccurate stock records,
- insurance and security costs.

These costs are particularly important in times of high interest rates, the wasted money (working capital) tied up in stock is seen in terms of the interest on debts which could have been paid off; or might not have been incurred in the first place had less money been so tied up.

Small batches, stored for very short periods, would tend to keep such costs to a minimum.

Transaction costs

These are costs associated with ordering, making and transporting the batch, and will include:

- ordering, invoicing, quality checking and payment for a batch,
- conveying the batch back and forth from one part of the factory to another for processing,
- transport of raw materials and components from suppliers,
- for items that are processed internally, rather than purchased from a component supplier, there will also be costs of tooling up and set-up time for the machinery used in processing the new batch, production planning, progress chasing etc.

Such costs will tend to be minimized by producing a small number of very large batches. The batch size which minimizes the sum of these transactional and storage costs is commonly referred to as the Economic Batch Quantity (EBQ) when deciding batch sizes for a production run; or Economic Order Quantity (EOQ) when arriving at an optimum order size for goods inwards. In many respects, the calculations used are the same for EBQ and EOQ, so we will only examine EOQ here.

The Economic Order Quantity

Since storage costs will be minimized by small batches, and transaction costs minimized by large batches, there is an implication of a trade-off. In practice, when these two costs are added together for a range of outputs, the cost graph tends to be crossover, as shown in Figure 4.2.

In an industrial situation it is only necessary to estimate the total cost function, and then use calculus to find the minimum cost output. Alternatively, much the same result could be obtained by simply graphing the function and reading off the minimum.

The total cost function is relatively easy to derive, using the following notation:

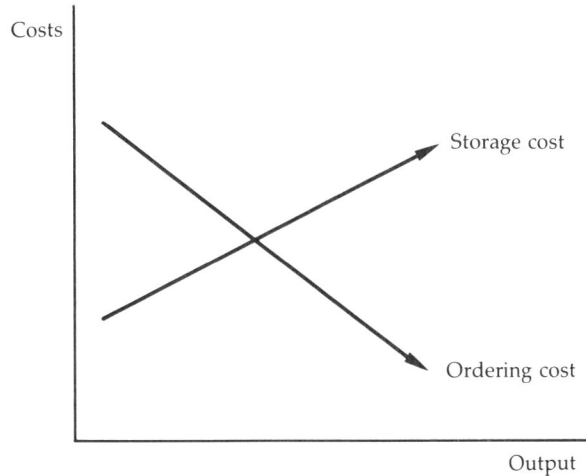

Figure 4.2 Economic
Order Quantity

$\overset{*}{Q}$ = Quantity needed

$\overset{*}{k}$ = Cost of holding an item in stock for a year

$\overset{*}{a}$ = Cost of placing an order

C = Total cost of ordering and carrying stock

D = Number of items ordered in each batch

T = Transaction costs

S = Storage costs

Terms denoted by a star indicate that a value is known, or can be estimated by the production manager. Costs and quantities refer to a consistent time period, normally one year.

Total cost = Transaction costs + Storage costs

or

1    $C = T+S$

Transaction costs will be the cost of making an order, multiplied by the number of orders made. The number of orders made can be estimated in advance by dividing expected sales by the number of items ordered (or made) in each batch.

2    $T = a\overset{*}{Q}/D$

Storage costs

Total storage costs will equal the cost of storing one item, multiplied by the average stock level. The average stock level is likely to be half the quantity delivered (or made) in each batch.

3    $S = \overset{*}{k}D/2$

Total cost therefore, is the sum of equations 2 and 3

4    $C = aQ/D + kD/2$ ** *

Since there is only one unknown on the right-hand side, the value for D, it is possible to graph a solution by trying various values for D.

However, the same solution can be found more swiftly, and much more elegantly, by taking the first derivative of total cost with respect to the size of each batch, and thus locating the output (D) which minimizes cost.

$$\frac{dC}{dD} = 0$$

So EBQ $= \sqrt{(2aQ/k)}$ * * *

Example: An electrical engineering company uses 25,000 standardized microchips per year. The cost of making an order to the supplier is some £400, and the cost of holding one microchip in stock for a year is estimated at £1. How many microchips should the supplier order in each batch?

EBQ $= \sqrt{(2 \times 400 \times 25,000/1)}$

$= 4472$ microchips per order

In real life, of course, the order quantity would need to conform to the package size offered by the seller, but assuming that the chips can be purchased in lots of 500, we can say that this producer will take six deliveries per year (one every two months) of around 4,500 chips per delivery.

This model is a little simple in that it ignores the likelihood of discounts for larger volume orders, and the effects of holding a minimum stock quantity. It further assumes that all of a batch is received at one moment in time, whereas a company which makes the components itself could replace these items gradually. Inclusion of these factors into the equation, where they make a difference, does not present a particular difficulty to the production planner. Moreover, the planner should ensure that costs have been minimized, rather than maximized. However, these complications will not be explored further here. After this brief look at cost planning and optimization, we turn to production planning.

Drawings trigger production

Once the batch sizes have been estimated, it is possible to move on and deal with production scheduling, materials requisition and any training necessary before production can begin. As suggested in the opening chapter, 'nothing happens on the shop floor without a drawing'; and engineers are therefore the key trigger to the production process. They produce bills of material as part of the design process, which is a list of every component needed, be it manufactured in-house or procured from an outside supplier. Job planners have to analyze the material procurement aspects for the purchasing department to organize, and identify those components for internal manufacture that will need new tooling and/or welding jigs. Moreover, planners will devise any new machining processes that may be necessary, as well as plan the methods of sub-assembly and final assembly which follow.

There are many routine logistics, machinery processes and job planning tasks, such as jigs and fixtures to be designed and made as important aids to reducing production time and hence costs. Traditionally, this was seen as purely a production function, albeit that engineers were employed here.

Control techniques

Although it may be possible to prepare entirely consistent and achievable production plans, it is rarely the case that the plans work without mishap. Illness, machine breakdown, inadequate quality control, clerical errors and revised sales estimates will inevitably cause deviations from the original schedule. In such a complex system, it is vital that production management both minimize the occurrence of these events and the wastage so caused, but also retain an accurate picture of where each batch is currently located, and its current stage of completion. Accountants will also require information on which to base their calculations on work in progress and cash flow forecasts, and the sales force need production information in order that they can quote realistic delivery dates and prices for special features.

A production control system is necessary to provide this information. Some typical production and control functions are locatable in Figure 4.3 and the associated tasks are described below.

Progress chasers

People are necessary to iron the kinks and bottlenecks out of cumbersome inflexible operations that develop. It can be quicker to manually by-pass the system and second guess the production plan, by direct negotiation with a foreman or supervisor, than put a cumbersome bureaucratic system right. Examples of this are particularly common when prototyping a new product, or implementing rapid design changes.

Furthermore, complicated storage plans occasionally break down, components become misplaced or simply lost, and production schedules are not met. It is therefore necessary to have people spending their time

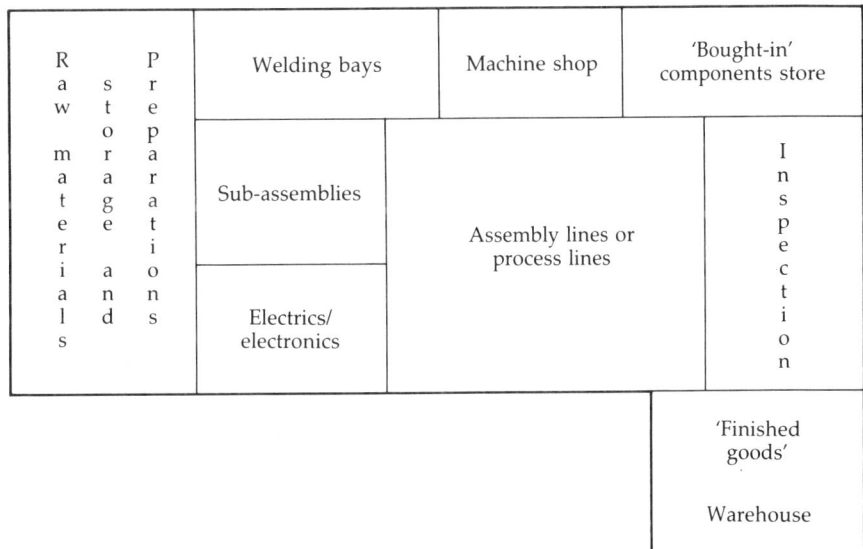

Figure 4.3 Traditional factory layout: C1970

finding out the location and condition of batches not found in the original location assigned to them by the production plan.

Inspection

The term 'an army of inspectors' can literally be true in large industries. At every stage of the production, goods inwards and finished goods processes can require inspection of some kind. The emphasis is on an inspector rejecting components that are not processed to a designated standard, rather than relying on the commonsense of the operators actually producing the goods. In the past, the onus has been on the customer inspecting the finished product on receipt to their premises, on a tacit buyer beware approach. Hence goods inwards inspection can be an important and expensive process, often causing delays if 100 per cent inspection of a batch was necessary, particularly if tight deadlines are being dictated for the finished product.

In such a situation, unacceptable short cuts may be taken, with suspect or unchecked components being used in the hope that they will work.

Designing and building jigs and fixtures

Any component made in batches, as well as many sub-assembly processes, render themselves to much shorter production times if the handling of the component is automated. Jigs and fixtures are handling structures, usually specially coloured for ready identification, which hold the component or assembly rigidly in place while operations are performed, so that human labour can concentrate on the actual operation itself. Be it welding, machining and assembling, jigs and fixtures make the job ergonomically efficient by saving labour time. This is where manufacturing effectiveness is prime in assuring competitive edge. The modern terms of 'manufacturing engineer' and 'manufacturing technology' depict the trend towards the fusion of these functions, to a closer proximity with the design function.

Work study, rate fixing and the unions

The planning of each job has to be broken down into 'standard labour times', which are set by work study specialists, to which standard rates of pay for each job are allocated. Performance over and above the standard is rewarded with bonus pay.

The discussion on work study can be a frequent cause of meetings between management and the various unions to sort out working and assessment methods and rates of pay. After 1977, The Health & Safety at Work Act increased the focus on ensuring safe working practices, as it made the company and staff more accountable for accidents and unsafe practices.

Materials handlers, crane and fork truck operators, and job pickers

The conveying of material and components between work stations and stores was a subject of demarcation focus up to and including the 1970s. 'Picking' the materials and components for each job can be seen as a skill in its own right, whereby 'pickers' service the work stations with parts picked to lists (called build schedules, with individual job numbers), from the planning department. These parts are then marshalled onto plinths for lifting to the job by forklift truck or crane. Since mistakes can be common, this is often an area of contention, with foremen having to intervene in the event of delays or mistaken component deliveries.

Stores and storekeepers, the inventory control system

The consequence of operating the Economic Ordering Quantity and Batch Quantity systems is a need for substantial storage space, often under cover, to avoid damage. Some companies segregate stores for production parts from spare parts stores for after-sales servicing, so that priorities between the production process and the crucial after-sales process do not clash.

Unless the company is operating a perfect system, where bought-in components arrive in the factory immediately prior to the production process, both bought components and in-house produced components need to be stockpiled until used. This is generally done in metal containers (called bins), stored neatly in racks, with each one given a bin location number. This means that manufacturing premises need substantial storage space, and a system of identification and retrieval, together with manual or automatic conveying systems to:

1 Take components to their bin locations.
2 Retrieve components as needed against individual job numbers as they progress through the manufacturing system.
3 Locate the bin whenever 1 or 2 failed.
4 Restock the bin when empty, or start a new bin if the old one can't be found.

A constant second guessing process can develop, as supervisors try to assess the stock level which could be reached, before it becomes necessary to restock the bins with a further economic batch quantity.

Considerable improvements became possible, in theory at least, with first and second generation IT equipment, such as the well known MRP I and MRP II systems on offer from IBM. MRP stands for 'material requisition (or requirements) planning'.

Grinders, welders, machinists and fitters, mechanical and electrical direct labour

These functions are the 'direct labour' content of the operation, in that their effort is accounted for as productive labour, as opposed to the production support staff (or indirect labour), discussed above. Direct labour is attributed to individual jobs by allocating each hour of labour to the job number traditionally recorded on a job card, but nowadays the record of hours, (and often minutes) can nearly always be computerized, with a PC on the shopfloor. Supervision is usually carried out by section leaders, called foremen, each controlling two or more charge-hands, who are the first filter for checking efficiency on the job, ironing out drawing interpretations and enforcing quality control by responding to quality reports from the inspection department. A daily production meeting is essential, to ensure smooth job scheduling, optimum material control and provisioning.

Western and Japanese production methods contrasted

Western methods evolved at a faster pace after the Second World War, based around the idea of large production volumes, and increasingly complex manufacturing control systems designed to keep a plant operational. However, by the 1970s, it was clear that Japanese companies were gaining a substantial manufacturing advantage in several industries, including automotive, electronics and robotic sectors. In 1986 Professor Parnaby of Lucas, published a series of comparisons between British and Japanese production systems:

	Japan	Britain
Ratio of support workers to production workers	0.5	1.4
Time between order and delivery (index)	70	100
Sales per employee (or value added)	£90,000	£30,000
Product cost (index)	70%	100%
Percentage delivery on time	95%	75%

These measurements are averages, and do not imply that excellent British companies of world standing did not exist. The reasons that might lie behind these rather depressing statistics are discussed below, and it is important to note that many of these discussions became dominant political arguments in the 1980s.

Production and support staff

In Britain's manufacturing industries, there were almost one and a half times more people involved in functions whose output could not be directly attributed to producing the product. These were supporting roles, called indirect labour or staff, including all those controlling and planning the production process as well as marketers, accountants, administrators etc. Those actually putting productive labour into making the product were the minority by a factor of three. This may reflect a British tendency of hierarchical over-control and over-measuring. The ratio may also reflect an effect of trade unionism, whereby strict functional demarcation, both on and off the shop floor, contributed to the well publicized poor productivity of the 1970s and early 1980s. For example, a British welder could not drive a fork-lift truck to collect and martial his own material; it needed a separate operator for this. Such a system often resulted in queuing and waiting for service. The implication is that the system was paradoxically top and bottom heavy. There is no implication that individuals were lazy, nor that there was no need for support staff and controllers. It is simply that poor structure meant that a great deal of hard work and clever thinking was misdirected. In contrast, the Japanese direct worker, on average, was supported by half an administrator, or almost three times less administrative support than the British equivalent carried.

Delivery times

We have already suggested that shorter delivery times are an important part of the augmented product, and thus represent a significant marketing tool.

The statistic on lead times, or delivery performance to the customer, is self explanatory. It implies that Japanese industrialists had an ability, on average, to deliver consistently 30 per cent sooner than their equivalents in British industry.

Sales per employee: the measure of value added

This is a crucial indicator of 'competitive edge', and can be looked at in two ways:

Each Japanese employee was producing three times as much revenue as his or her equivalent British competitor, or

The equivalent Japanese company, turning over the same revenue and producing equivalent products, could do so employing only one-third of the people required by its equivalent British competitor.

As it is impossible for people to work three times as fast on a straight pro rata manual basis, even in sweat-shop environments, it is obvious that the Japanese were paying incisive and radical attention to their methods of working. This involved an ongoing look at motivation and teamwork and, arguably, creating an environment of trust by guaranteeing lifetime employment to a majority of workers, while simultaneously investing in automation processes. In many Japanese industries, major risks were taken in capital investment for productive machinery and robotics for component handling processes. The results show up in the productivity comparisons above; not only were the processes more productive, but they also provided an additional bonus of consistently higher quality. But the biggest shock of all to competitors was the growing Japanese legend that they were consistently more flexible in meeting the rapidly changing product demands of sophisticated Western markets.

Product cost

In many industries, Japanese producers had achieved significant cost advantages, in some cases as much as 30 per cent lower cost than a UK competitor. This cost leadership gave an enormous competitive advantage when forming strategic and tactical marketing plans. Matching a higher cost competitor's price in a boom period will generate higher profits for the low cost company, allowing greater reinvestment in plant and product development. In a recession price wars inevitably start. Clearly lowest cost ceilings give the necessary resilience to last out longer than one's competitors, while still generating some profit. Companies with higher costs can be forced to operate at a loss, eventually going bankrupt, and leaving a vacuum in the market for the survivors to exploit.

In the 1970s, there was still a belief among Western business that the Japanese were employing suicidal low pricing policies to enter markets and establish a large market share, and that this was a short-term strategy which could not be sustained. In fact these Japanese companies were still trading profitably while apparently giving 10 to 20 per cent more economic value to the customer than their Western competitors, making them well placed to win global markets.

Meeting delivery promises

In addition to offering faster delivery times when contracts were made, Japanese credibility soared as it was realized that their word was their bond in this respect. Stories abounded of unusual methods being employed to ensure customer satisfaction on delivery and service back-up, be it the hiring of a jumbo jet to beat a deadline, or compensation in lieu of the product in the case of dissatisfaction. The customer was made to feel important. The Japanese were more reliable on delivery promises by a consistent 20 per cent in the British home market, despite offering shorter delivery times from a source on the other side of the world.

By way of contrast, one major British exporter of paper products would generate three forms at the time of a contract; the 'Promise Note', which confirmed the agreed delivery date, a 'Re-Promise Note', which would be sent to advise the customer of a new delivery date after the agreed date had passed, and a 'Second Promise Note', which would advise the customer of a third delivery date after the second had passed. The re-promise note would be used in most cases, and use of the second promise note was by no means uncommon.

The Japanese methods explored: the early 1980s

It was soon realized that this consistent performance by most Japanese companies, supplying products to Britain, was no fluke. They soon established a reputation for unprecedented standards of reliability, notably in the car industry. It was not understood how such superior quality could be achieved at such competitive cost levels; after all, for equivalent metal content, similar engine sizes, but superior electronics gadgetry, Japanese cars were gaining commendations (for instance from *Which?* magazine), for consistent excellence on performance during warranty periods, at unbeatable retail prices (despite heavy import tariffs).

A similar scenario was developing in North America, despite their average indigenous car prices being a half of the European average. Urgent competitive analysis and a major rethink were necessary by industrialists all over the western world. Such leading industrialists as Sam Toy, then Chairman of Fords of Britain, visited Japan themselves to learn, at first hand, how these results had been achieved. Toy was later to use his visit as the basis for implementing a radically new manufacturing policy at Dagenham called 'After Japan'.

The delegation from Ford, like other such delegations, found the virtual non-existence of inspectors on Japanese production lines, but rather an emphasis on getting it right first time, fostered by trust in the workforce to be their own best critics. Management would often support, rather than control, initiatives from the workforce, who would meet in 'quality circles' to continually review and improve operations. Problems on the production line would be reviewed and resolved by team leaders on the shopfloor, reference to management was seldom required. Many of these practices were quickly adopted by Ford as part of their new strategy.

They also recognized that excess inventory, rather than acting as buffer safeguards, hides the very flaws which need controlling by management. Further, the reasons for delays and bottlenecks are obscured by the very complexity of the systems designed to prevent production delays.

They also saw at first hand the Japanese ability to deal with the growing nightmare scenario for Western industry, namely the increasing demand by discerning consumers for a wider variety of product features. Consumers have continued to demand more and more specialized products; for example the simple plimsoll has been displaced by an enormous variety of training shoes designed for different sports, meeting the peculiarities of individual running styles and colour/design fashions.

A similar explosion of variation is seen in the car market, with manufacturers now being forced to develop a bewildering variety of upmarket versions based around a standard model, e.g. Fords XR2, 3 and 4; and, even at the budget end of the market, a wide variety of Fiat Pandas. This demonstrates how consumer pressures have shortened product life cycle durations, from a stable ten years or so (typical of the 1970s), to two or three years at present, before product rejuvenation or complete design overhaul is needed, in the now axiomatic need to differentiate and keep ahead of competition.

The demands on manufacturing methods flexible enough to cope with this continual change, such as a widening of product ranges, has led, for a batch producer, to shorter and shorter production runs and escalating costs. The Japanese, far from fearing greater variety, embraced it as a powerful marketing tool and developed systems for shortening production runs without significantly increasing costs. Sobering factors for those British companies who were to survive the recession of the early 1980s, only to face immediately a tougher business environment, with the consumer as king, and spoilt for choice by Japanese innovations. The need for a strategic manufacturing outlook was gathering speed.

The subsequent sea change in the Western business world was far-reaching and painful, and by the mid-1980s Britain saw a fairly decisive swing to service oriented and screwdriver industries. Many manufacturing and engineering companies disappeared during the recession, but, argued some commentators, the remaining 'fitter and leaner' industrial companies were in a position to achieve significant productivity gains by sustained investment in automation, changes in working practices and improved organizational and managerial systems.

Today, newspapers are full of statistics and case studies on how well the Japanese-owned (and influenced) car industry in Britain is now doing in terms of quality, and their contribution to rejuvenating that industry's export drive to Europe: Toyota and Honda are building factories in Derbyshire and Swindon respectively to meet forecast increases in demand for their models by the European market over the next ten years. The success of Nissan, and Rover's joint venture with Honda are well documented. The moribund Bedford company, once famous for its vans, now works with Isuzu using Japanese style methods which have revitalized the company, now trading as IBC.

Further reading

More on basic production methods can be found in the relevant section of *Management in Theory and Practice*, by G.H. Cole (DPP Publications, 1989), while *Production/Operations Management*, by R. Schmenner (Macmillan, 1990) is a more focused operations manual. Developments in manufacturing industries are well traced in *Manufacturing Industry since 1870*, by M. Ackrill (Phillip Allen, 1987). Reasons for the relative decline of UK manufacturing can be looked at by examining specific business case studies in *The SuperChiefs*, by R. Heller (Mercury, 1992). Other related topics are explored in *The Economy under Mrs Thatcher, 1979–1990*, by C. Johnson (Penguin, 1990), and M.J. Wiener's *English Culture and the Decline of the Industrial Spirit, 1850–1980* (Penguin, 1981), although these books do not explicitly discuss manufacturing methods.

Tutorial exercises

4.1 A distributor of industrial products in the Middle East has just begun trading in electronic mousetraps. Early market research suggests that he will sell around 5,000 mousetraps in a year (spread equally over the year). Each mousetrap will cost $10 in delivery charges, and each order he makes will cost him $150 to process. The cost of holding one mousetrap in stock is $1.50 per year. The distributor intends to follow his usual custom of ordering one-twelfth of his expected annual sales each month.

(a) Advise the distributor on the wisdom (or otherwise) of his intended technique for choosing the size and frequency of his orders of electronic mousetraps.

(b) What size and frequency of order would you suggest?

(c) Comment critically on any weaknesses of the method you have used in (b).

4.2 Dispute at Sticky Sweets.

Sticky Sweets is a small company primarily concerned with making children's sweets, although it has sidelines in specialist chocolate and also runs a small confectionery shop in Leicester. The business – still owned and managed by Mr Sticky – had chugged along for some time in an unspectacular way, but he felt the business had much greater potential.

Mr Sticky decided on a major expansion plan intended to make Sticky Sweets a major brand name. He took on extra sales personnel and modernized his plant so that he could produce larger batches more efficiently.

The first part of his plan went well, new plant was installed and became operational very quickly. He anticipated that output could be 40 per cent higher if he could run the new plant at full capacity. His newly recruited sales team performed beyond his expectations, and won sufficient orders for Mr Sticky to run the plant at full capacity.

The first sign that all was not well was the notable reduction in stock levels of finished goods in stock as sales volumes increased, despite the new plant which had been installed to provide extra capacity. Mr Sticky realized that the actual output of his plant was substantially lower than he should have been able to expect, and stocks of finished goods were being run down to make up the difference between sales promised and actual output, while stores of raw materials rose rapidly. Further examination showed that output had been dramatically improved by the new technology in some parts of the factory, but had remained at old levels in others, creating bottlenecks on the production line and large stocks of part processed sweets to build up.

His first attempt at solving the problem had some beneficial impact. Productivity did improve when workers were properly trained to operate the new gadgets that had been installed, but even this failed to lift output and productivity per worker to levels sufficient to reverse the depletion of stock.

There were several departments which showed no sign of improvement; but most noticeable was the workshop which made toffee. The work of toffee making is extremely tedious, requiring several routine operations to cook, cool, solidify and flatten the toffee before it is taken away and wrapped in another building. The working conditions in the workshop were unpleasant; there was the constant heat and smell of boiling toffee and frequent minor burns caused by splashing toffee to put up with, and the noise from the continual ferrying of raw materials and cooked toffee by fork-lift truck made it difficult for workers to speak to each other. Staff turnover had been high for some time, and higher wages had not prevented workers from leaving whenever they had the opportunity. Mr Sticky placed a very experienced supervisor, with a reputation for driving his workers, in charge of the workshop, hoping to improve the performance of the section.

The increasing bottlenecks in those departments still producing at low levels of output began to cause cashflow problems as work in progress continued to increase. Mr Sticky began offering further bonus and overtime payments to workers in bottleneck departments like the toffee shop, to speed up the throughput, but with little effect. Workers in other parts of the factory began to resent the extra cash paid to workers in departments which had not raised their output.

Mr Sticky began to examine working practices and methods, and to consider the possibility of introducing shiftwork, when an unfortunate incident occurred. The workers in the toffee workshop, in which output had risen little and productivity not at all, discovered that the contract laundry had returned dirty overalls on a Monday morning, rather than clean ones. The toffee makers refused to wear the sweaty, smelly clothes (which had been in a steel trunk for a weekend). Unfortunately, their supervisor took a different view and threatened to sack workers who would not start work. After several insults from both sides, the toffee makers walked out on strike. Several other departments stopped work in solidarity. When clean overalls arrived in the afternoon, the factory resumed work – but Mr Sticky realized that there is little goodwill at present between his 45 workers and the management.

He believes he has around six weeks before his stocks of finished goods runs out. If he has not solved his problems by then, he will have to disappoint his new customers.

He has called you in as a consultant. He wants you to examine ways of improving the performance of the toffee shop, with a programme which can be transferred to other trouble spots. He would also welcome any other advice you may care to give on his expansion programme.

4.3 Quikcabin is a company which make a variety of temporary outbuildings and prefabricated sheds.

Quikcabin buy brass plated screws from a supplier to make their outbuildings. Although brass screws are twice as expensive as coated steel ones, they will last for twenty-five years, twice as long as the steel alternative. They actually need some 200,000 screws a year, and have got into the habit of buying two lots of 100,000 screws each year because the supplier offers a 5 per cent discount on orders of 100,000 screws or more.

It costs £5 to store 100 screws for a year. The supplier is a local merchant, so delivery costs are relatively small and deliveries can normally be made within a few days. Nonetheless, it still costs around £40 to make a single order. The trade price of the screws is £100 per 5,000 screws for orders under 100,000.

As part of the production process of prefabricated buildings at Quikcabin, some components have to be bonded by a particular process called mucilage. Output is normally between two and three units per day, which is currently enough to meet production schedules and often runs ahead. The work is very skilled and specialist, so it is not integrated into the main production line or process.

An industrial placement student has noticed that the costs in the mucilage shop seem to be related by the function:

$$C = 0.25 x^3 - 10x^2 + 50x$$

Where C in pounds (£) is the total cost of producing x items in one day.

This is checked out and, remarkably, does seem to be close to the truth. The range tested included very busy periods, when x was five or six units per day, down to the present level of two or occasionally three units per day.

(a) At what level of output would the mucilage shop produce most efficiently?

(b) The market price for bespoke pre-bonded components of equivalent quality is £43 each. Show how this information can be used? What action should be taken as a result of any calculations you may make?

(c) The production director is so excited at the results obtained from this work that he wishes to have the whole production process similarly analysed. What advice would you offer before the project starts?

(d) The stocks of brass screws have reached their reorder level. What advice would you give to the purchasing manager?

5 Modern production methods

It is now widely held that many of the Japanese inspired production methods are more suited than traditional Western methods for marketing to dynamic, competitive markets. In this chapter we examine the key elements of modern production methods. These ideas should be studied with the factory layout plan shown in Figure 5.1(a).

Flexible manufacturing systems

Individual machines are organized into groups called 'cells', that are so flexible that down time (also known as dead time) necessary for tooling change-overs has been reduced to minutes rather than hours. The mission here is a drive towards the ideal economic batch quantity of one. The layout of a cell is shown in Figure 5.1(b).

Imagine each cell as an integrated mini machining and sub-assembly unit where the traditional linear layout has been condensed into a circular workstation – so that instead of the component being conveyed sequentially to each operation, the various machines involved are now placed round the component. Minimum component conveying is now possible, and when needed, is often done via a robot, which ensures maximum precision. In reality, the flow process is complex, involving a number of cells which in their turn are integrated. The reason for the complexity is dictated by the frequent need for options demanded on top of the basic product; for example, the special trims, wheels and features which go to make up the XR3 and GTi ranges of cars, versus the standard L ranges. In practice, every industry evolves its own peculiarities which require careful production planning to cater for the plethora of variety demanded by the modern consumer.

Figure 5.1 (a) Modern factory layout; (b) Organization of cells around components

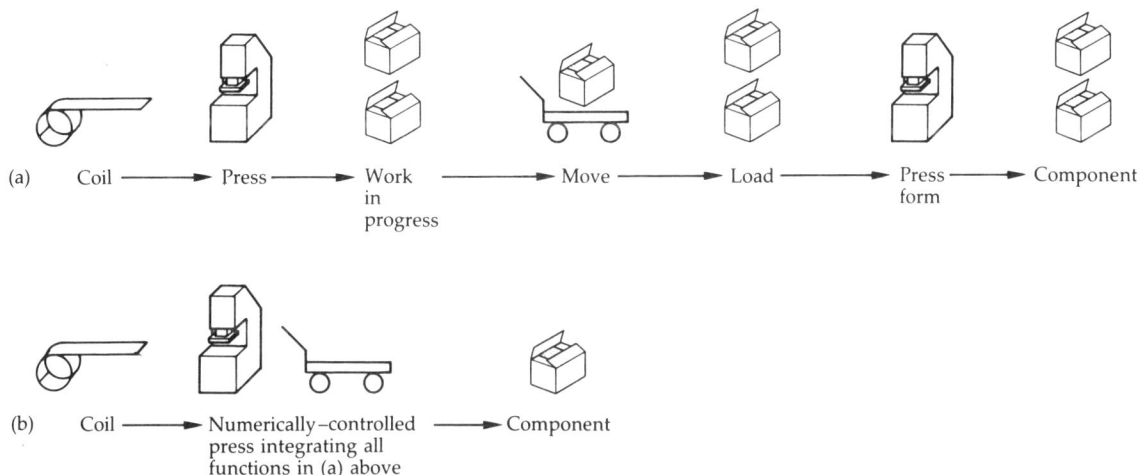

(a) Coil ⟶ Press ⟶ Work in progress ⟶ Move ⟶ Load ⟶ Press form ⟶ Component

(b) Coil ⟶ Numerically–controlled press integrating all functions in (a) above ⟶ Component

Design for manufacture

In order to manage these flexible systems, business managers, engineers and project managers must now be expert at designing innovative products which are easily makeable, and be completely aware of the capital investment implications of their product designs on manufacturing equipment and production systems. In other words, production processes must be considered at the design stage. Thus Design for Manufacture (DFM) and Design for Assembly (DFA) have become as important as the need to ensure that the product actually works. This places ever greater strain on the design skill involved in production.

Computer technology can do much to reduce the problems concerned. Computer Aided Design (CAD) enables the designer to produce 3D images of the finished product relatively quickly, using pre-existing components wherever possible. The software will identify these parts by outputting a parts list, greatly facilitating bills of materials for the buying function. The engineering drawing function can be incorporated by integrating CAD with Computer Aided Design and Draughting (CADD). CAD software can also generate numeric control codes which will operate automatic devices as part of a Computer Aided Manufacture (CAM) system. CAM systems are capable of setting and driving plant ranging from unsophisticated drilling machines to advanced laser technology and robotics. CAD and CAM used together are generally referred to as CAD/CAM.

As software design continues to develop, knowledge-based systems (or expert systems) are expected to do much to further improve DFM.

Traditional production methods could benefit from the use of CAD/CAM, but its use is most advantageous in Computer Integrated Manufacturing (CIM) systems, using FMS. Flexible manufacturing incorporates some of the remote control technology originally devised for the space programme. Machinery can be pre-programmed for production volumes, to accurately inspect its own output, and to automatically re-tool itself at the end of a run, using robotics to both marshall material and change a tool to pin-point accuracy. Networked and simply operable PCs on the shopfloor enable the production flows to be matched to receipt of orders without the need for layers of progress-chasing administrators. The integrated control system co-ordinates the operations automatically, at cell level.

Production can be further integrated by direct computerized links with key customers and suppliers. The communications and IT revolution, and Electronic Data Interchange particularly, have driven far-reaching changes on the shopfloor, both in machinery terms and in working philosophy.

Those entering a Japanese style manufacturing plant today may wonder why there was ever a need for material requisition planning (MRP) and process scheduling (which involves considerable administrative manpower and the commitment to spending unnecessary capital well in advance of anticipated orders). In some cases, in heavy engineering for instance, lead time commitments to suppliers could be many months ahead of delivering the end product, so that the cycle of capital tied up in advance of being paid by the customer could even be more than a year.

The increasingly turbulent business environment leads to ever greater competitive pressure to respond to changing consumer demands, and tightening financial constraints dictate the need to synchronize production quantities to the fickle ups and downs of buyer behaviour so that stock holding costs can be minimized. Traditional systems largely fail to meet these exacting demands.

Kanban pull systems

In a Japanese style production system, manufacturing processes are actually triggered by the customer, rather than a production plan based on anticipated sales. Actual orders generate small volumes of production, co-ordinated by a system referred to as Kanban; which, in its early days, involved the movement of an empty component container to the source of the component, which could be another production cell. If refilling the empty container caused the cell to empty another container of inputs, then this would be returned to its source, which could be another cell or a goods inwards section. Set out sequentially, the stages are:

1 Fill an empty container with designated parts predetermined at the job planning stage;
2 Deliver this container to its point of use where the component or part is needed – this could be at a machine, robot handling system or removed manually;
3 The emptying of the container automatically triggers the movement of a card with instructions to refill the container again with the designated component/part for that container;
4 This implies that two simple cards for each container control the whole system. One card triggers the 'making' of the part or component in its cell. The second card triggers the conveying of the container to the next cell 'up line' of the process. The empty container is then returned in 'the loop' with the first card, which triggers the whole process again. Should there be no demand subsequently to convey the container 'up line', the process stops until a customer order 'pulls' the system into operation again.

This operation is shown in Figure 5.2.

Figure 5.2 Kanban pull system

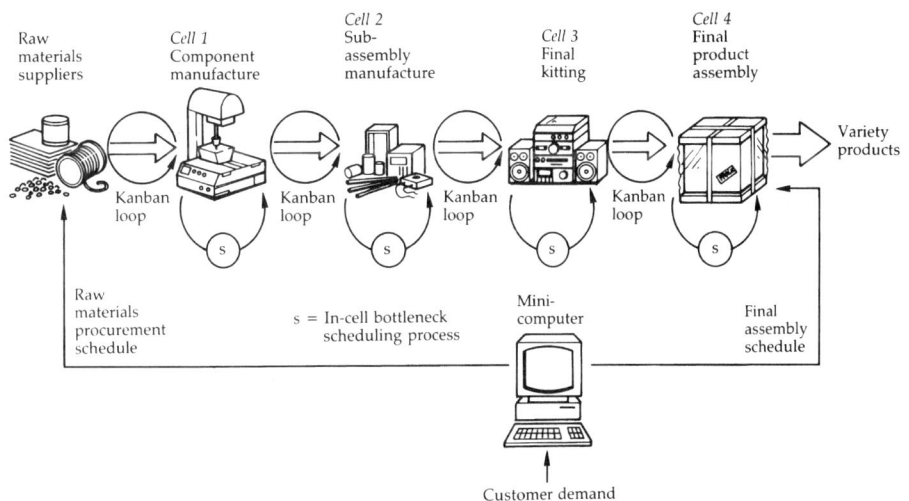

93

The philosophy behind the containers is to ensure that only what is needed is produced. If the containers are full, no more parts or components are produced until containers are again empty. Cell workers then attend to such activities as monitoring daily throughput of parts and components, re-ordering raw materials and components, checking quality control data and maintaining the machinery. In the 1990s, the cards and containers have long been replaced by a PC which logs output, updates stock records (and prompts an order if appropriate) and triggers further production automatically.

This contrasts with traditional Western methods of guessing sales in advance, and using this guess as a basis for materials purchasing and production planning. Kanban, the brainchild of Toyota, is referred to as a pull system, depicting that it responds purely to actual orders received, sometimes by the hour. The MRP system, evolving in the West since the Industrial Revolution, is designated a push system, as the process implies.

Push and pull systems combined

As each business is unique, many hybrid systems have evolved, and manufacturing technology is an industry in its own right which has fuelled the IT revolution among the major multinationals. Some companies have opted outright for the total Japanese approach which relies on stand-alone first generation networked PCs. Other companies, particularly in bespoke engineering markets, such as aircraft engines and power generation, where small lot sizes are the norm, need to employ a hybrid system, which is likely to be a combination of both a push and a pull system. The IT systems in this case are likely to be of second generation sophistication, involving substantial guess-working on material needs, to cater for supply chains not yet used to operating JIT, and meeting the needs dictated by industries who do not have the advantages of high volume. Black and Decker, for example, use JIT for their 100 most common components and MRP for the rest.

The engineer has additional challenges to design components suitable for a Kanban system, and build innovative products from such components.

Minimum stock levels

We have already examined the notion of Economic Batch Quantity, and pointed out the need to minimize the number of machine set-ups, and the way this generates large stock levels (also called inventory levels).

These high stock levels often hide inferior quality or even unusable batches, for many months, delaying product improvements, causing major storage overheads and tying up considerable amounts of capital unnecessarily. For example, in 1981 one small plastic moulding company made plastic cases for the Sinclair microcomputer for two months before the buyer received the first batch; he then noticed a flaw which made the whole batch useless. Fortunately, the company was able to salvage the material used, but all the machine time, labour time and storage costs were wasted. Moreover, the company had to run an emergency overtime schedule in order to make delivery by the agreed date, causing added expense and disrupting production schedules elsewhere.

The Japanese alternative

After the Second World War, Japanese industrialists had neither the capital or materials to run a high inventory system, and were forced to develop alternatives. Clearly Kanban could do much to reduce inventory levels indirectly, but they further resolved to develop a new system, with a new means of production, now called 'Just in Time', (JIT), which, in theory at least, was designed to reduce the economic order quantity to one unit.

The philosophy of JIT involves designing new machinery and handling systems that eliminate the need for planning in advance to cater for the shortcomings of inflexible machinery and production planning. Planning for production by JIT involves consideration of production methods at product design stages, that is, the product must be designed so that it can be made quickly using standard, off-the-shelf components and standardized production methods. The requirement for standardized components will include even the very first stages in the company's supply of components, from the choice of nuts and bolts (often two stages further back in the supply chain), through to production processing, testing and packaging. Design will also encompass the whole life-cycle of a product.

The initial design must allow flexibility, both in the product itself and the production method, to add features and competitive enhancements to the product at any stage from inception to prototype testing and market launch, through to maturity at the end of its life cycle two to five years hence. This involves multi-disciplined project management, where the methods of making the new product are taken into account at the design stage, and where the potential disruption on the shopfloor, and production costs of the inevitable optional extra product features, can be forecast.

JIT also involves both deliveries to customers and from suppliers. The goal is to develop the ability to respond with a 'same day service' without producing buffer stock in advance of orders from customers, and the ability to cope with a variety of specifications without prejudicing the speed of delivery. JIT does not require that suppliers hold large inventories of components, the intention is that suppliers also turn to JIT production.

Introducing JIT

Many companies opt for trying JIT out on a pilot basis, using a hand-picked project team to kick-start the process. Careful monitoring then follows, while the basic concepts are adapted to the unique needs of each company. Trade-off decisions are always required between what constitutes a 'volume line', justifying JIT treatment, and what constitutes a smaller process best suited to batch production.

Management will often apply Pareto analysis, which is better known as the 80:20 Rule, introduced in a marketing context in Chapter 3. Pareto analysis would be used in a production context to help identify ways of simplifying process flows, even in complex bespoke industries, to a half dozen clearly identifiable 'sub businesses', with well defined volume and variety characteristics. A Pareto analysis applied to a manufacturing company would be likely to reveal the following kinds of results:

1 Eighty per cent of the cost of sales value of a product are typically in 20 per cent of its components.

This gives important and immediate focus on materials management decisions making, as there are only 20 per cent by 'volume' of high value parts needing close and continuous scrutiny, and the number of key suppliers needing this attention is far more manageable than having to monitor every supplier with the same degree of attention.

2 Eighty per cent of the company's revenue will be typically generated by 20 per cent of its product ranges.

These would normally correlate to the company's high volume or standard product lines, and both the volume and cost of sales value will be substantial.

Putting 1 and 2 together would identify parts which were expensive and used extensively in the manufacture of the company's most profitable product lines. One, or several, of these parts could be managed on JIT lines as a pilot project, and would be able to demonstrate immediate and substantial cost benefits if successful.

The same analysis would also highlight components which would not be switched to JIT because the savings would never recover the initial investment. As was emphasized in the previous chapter, manufacturing processes are a combination of bespoke, batch and continuous processes in reality. JIT has displaced batch production to a considerable extent, but is unlikely to replace it completely. It is common to speak of 'A' lines, those immediately suitable for JIT production, 'B' lines where JIT might be implemented at some stage and 'C' lines, where JIT would seem to be a non-starter.

JIT clearly confers great marketing benefits in terms of flexibility and delivery times, and has a profound impact on a company's finances. The latter is discussed in some detail in Chapter 10. Further, it enables manufacturers to cope with increasing product diversity, as the focus on reducing costs for short production runs will enable small market segments to be targeted profitably, and a greater variety of products to be offered within each segment (Figure 5.3). However, JIT will not work without dedicated commitment to the philosophy once the pilot projects have proved their worth.

Figure 5.3 How the JIT process caters for variety

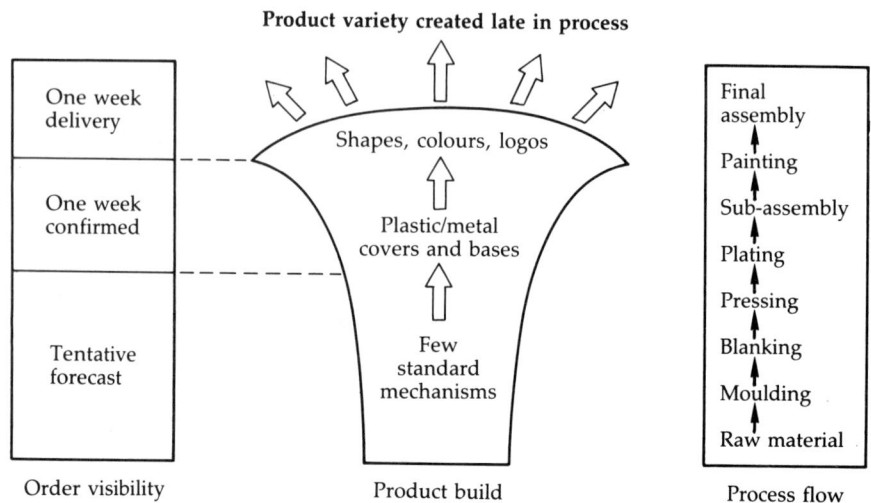

Product variety created late in process

One week delivery / One week confirmed / Tentative forecast

Shapes, colours, logos

Plastic/metal covers and bases

Few standard mechanisms

Final assembly / Painting / Sub-assembly / Plating / Pressing / Blanking / Moulding / Raw material

Order visibility   Product build   Process flow

Problems with JIT

The idea of JIT has revolutionized the way in which production is planned. But changing from traditional methods to JIT is extremely fraught, requiring new working practices, new technology, training and a radical change to the way managers plan production activity. During the introduction of JIT, production performance may well decline dramatically, and major investment will be needed in new and radically different machinery. One should not underestimate the powerful people issues involved as, understandably, suspicion of change can become a powerful demotivator. Invariably, labour turnover rises significantly as new training and working methods are introduced. Further, the changeover must presumably take place while the traditional business still operates, unless, as can sometimes be inevitable, a move to a new purpose-built site is not more cost effective. Teething problems can typically take three years to resolve. The topic of change management is explored further in Chapter 10 and 12.

Contrasting the before and after diagrams on pages 81 and 101, it is easy to imagine that such a change is obvious and desirable. But it should not be assumed that JIT is always easier to manage than traditional methods. Indeed, it can be argued that one of the strengths of JIT is that it highlights many problems, such as poor planning and variable quality, which traditional methods are deemed to be extremely good at disguising. Highlighting these underlying problems is pointless unless there is also considerable commitment to resolving these difficulties as they arise. Many managers and operatives who are experiencing the transition at present are far from convinced that there will be, ultimately, an improvement that will justify the efforts made.

A commitment to quality

So far this chapter has concentrated on production methods; but it is equally important to outline the growing importance of quality. Traditionally, quality control worked on the principle of an acceptable failure rate. It was calculated that the cost of detecting and preventing poor quality was, above a certain level, greater than any benefits derived.

However, many Japanese companies evolved a quality principle which not only required zero defects, but further stated that this standard should be achieved 'first time', meaning that adjustments and re-processing of components should never be necessary. Initially, such objectives were not taken seriously outside Japan, but as the performance and quality of Japanese products improved, buyers began to raise their expectations. Specifications of 'acceptable' failure began to toughen. The paper exporter referred to earlier in this chapter found by the mid 1970s, when they first began to test the quality of their products, that their production processes and equipment (which had been installed in the 1930s) were incapable of producing to specifications set by their major purchasers.

More recently, managers have moved towards adopting the philosophy of total quality management (TQM). It is the subject of many textbooks in its own right. It is a disciplined, all-encompassing philosophy which starts with total commitment from top management downwards to perform all functions to a predetermined and measurable level. The criteria are generally customer orientated, although customer is defined more widely than in its marketing context.

External customers are those who use or buy the organization's products and services. The dimensional or performance criteria for a manufactured product might be a customer reject level of less than 0.001 per cent or better in any batch delivery of more than 2000 units. Similarly a county council might seek to guarantee a meaningful answer to any query within 24 hours of the call. The Department of Social Security lays down guidelines for the maximum time which should elapse between claims and receipt of certain benefits (although county councils and the DSS may have some way to go before they reach all their desired levels of achievement). In other words, the level of quality is specified down to its constituent measurable components, both quantitatively and qualitatively.

In manufacturing, with the virtual elimination of inspection departments, the focus and need for automatic dimensional monitoring devices are recognized. A whole new industry, statistical process control (SPC) has sprung up. Just as the name implies, this is the continuous dimensional monitoring of critical components and sub-assemblies using electronic eyes to record and analyse, along well known statistical procedures. Any deviation from various specified standards set for the machine is identified immediately. Corrective action is also well defined, so that the production process is not interrupted.

However, TQM refers to much more than standards for the final product. In every operational area of the business, quality is an issue which should involve staff at all levels and in all functions. At operational levels, project teams, referred to as quality circles, instil a disciplined teamwork approach to quality in production and administration processes directly. Quality circles may be staffed by volunteers, their prime aim being continuously to improve the organization and delivery of the tasks they are required to carry out. These improvements can seem trivial when taken individually, but when added to thousands of other minute improvements (together with a few dozen major improvements hopefully) they can make great differences to operational efficiency, quality improvements and cost savings.

Quality circles operate by looking at each stage of every production and administrative process. Staff are required to identify their internal customers, and seek to improve the service to those customers. For a trivial, but real example, an engineer might identify a secretary as a customer, since the secretary is required to receive the engineer's scribbled notes as an input and generate formal letters and memoranda as an output. An improvement in quality would undoubtedly result if the engineer could be prevailed upon to make written communications legible. It is likely that the work of the engineer could be improved if the secretary were to adjust his or her work in some way. A similar benefit could be obtained if finance clerks were to complete expense claim forms on behalf of sales staff, rather than return the forms with complaints when mistakes are made. Again, no doubt sales staff could improve the work of the accounts department by submitting claims on time with the appropriate receipts.

A process analysed in this way is referred to as a quality chain, and is clearly beneficial to all involved in the chain, and can generate a never ending series of improvements. The major criticism levelled at the system is that preoccupation with internal customers may deflect attention from satisfying the requirements of external customers, who, after all, finance the whole chain by purchasing the end product.

Management may support quality initiatives with motivational competitions, aimed at beating current best performances, listening to and acting on suggestions where changing an established practice or capital investment in machinery will achieve better performance. However, refusal to support quality circles by resourcing their innovations (or failing to give an adequate explanation for such lack of support) will quickly lead to their demise.

SPC, and the fusion of materials management with quality control at board level, and the small but growing number of quality circles are evidence that poor quality is no longer accepted. It is now taken for granted in white goods and automotive markets, with, for instance, the trend for extended warranty periods, sometimes to a level almost three-fold traditional norms. The constant toughening of product liability laws and the political focus on performance charters put quality prominently on the strategic agenda. Clearly, passing such operational control to the workforce requires a radical change in managerial attitude in most cases, and is a world apart from quality derived from the traditional suggestion box.

The TQM philosophy is now adopted by many serious global manufacturers, and by national quality institutions, as a 'way of doing business'. A commitment to quality has major marketing benefits, as discussed in Chapter 2. Moreover, a high quality standard is necessary for the effective operation of JIT, since components are to be delivered to the assembly station 'just in time', often being on site for only the previous hour. In fact, daily and hourly component deliveries are now very common in volume industries. The delays which internal inspection, rejection, repair and replacement would cause on the process would now be deemed unacceptable, and would add unnecessary cost to the product. Instead, the supplier is responsible for controlling quality performance to negotiated predetermined standards. This implies a change in stance from the traditional adversarial treatment of suppliers, to partnership approaches, where trust is fostered through the agreed monitoring and implementation of long-term supply contracts.

Quality standards

In Britain, the British Standard BS5750 is now a minimum prerequisite for an acceptable quality rating, and some industrial buyers will only trade with BS5750 accredited suppliers. This encompasses all aspects of the way a company runs, ranging from its procedures, paperwork and IT systems to manufacturing and monitoring processes. Communication is assessed at inter-departmental level, as well as at the customer sales and service interfaces. However, the system has been criticized for generating further bureaucratic monitoring and control processes which add little value to the product or service received by the consumer, and falls short of other quality standards offered abroad.

Figure 5.4 Complexity of component flow lines in a traditional production layout

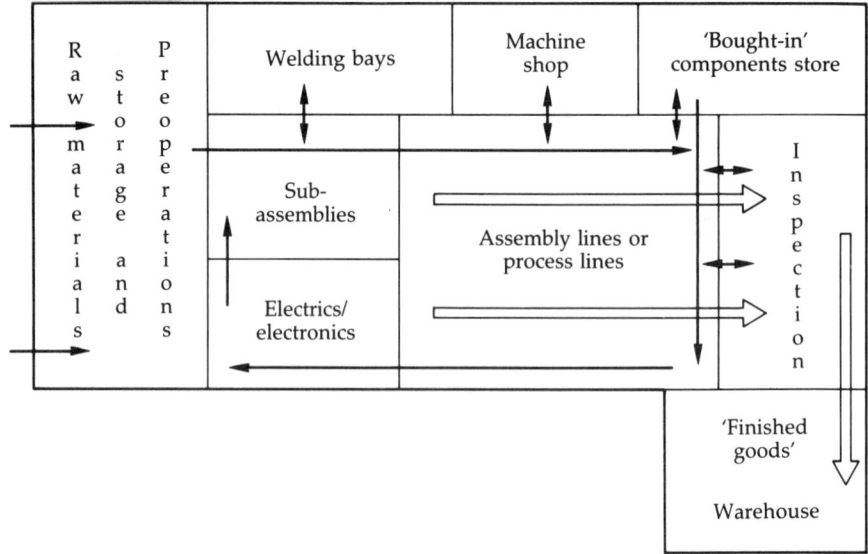

Modern production methods on the shop floor

For this section, you should look again at the layout of a traditional factory, first shown in Figure 4.3 on page 181, but reproduced in Figure 5.4 with material handling routes sketched in. Successful factories like this still exist today, although the pressures to be globally competitive mean that most management teams have to consider far-reaching change of some description on the shop floor. If it is not the adoption of JIT, or a half dozen other modern production processes such as OPT, MRPII etc. already described in Chapter 4, it almost certainly will already have been a radical change in quality control, both in methodology and meteorological equipment. Quality is an all pervasive function from which no company employee escapes, and as such everybody is accountable to the extent that dedicated inspectors are fast being absorbed and integrated into prime manufacturing functions.

Figure 5.4 depicts the complex material handling routes dictated by a traditional factory floor layout. The need to take components 'to the job' dictates handling – by personal or mechanical means, none of which adds value to the product. Chapter 4 covered these limitations in detail: suffice to say that any layout which minimizes the frequency of handling, distance of component conveyance and extent of dead time that components experience before being incorporated into the product is worth looking at. However, once the focus starts, it is often realized that tinkering at the edges of the real problem – namely the innate inefficiency of traditional factory layouts is a process of diminishing returns: radicalism is often the only answer to establishing competitive advantage.

Figure 5.5 shows those parts of the production floor most affected by JIT:

1 The shaded area shows how much storage space can become redundant by finding suppliers prepared to operate to the company's JIT requirements. The limited space is still needed to cover for the special materials

Welding bays   Machine shop   'Bought-in' components store

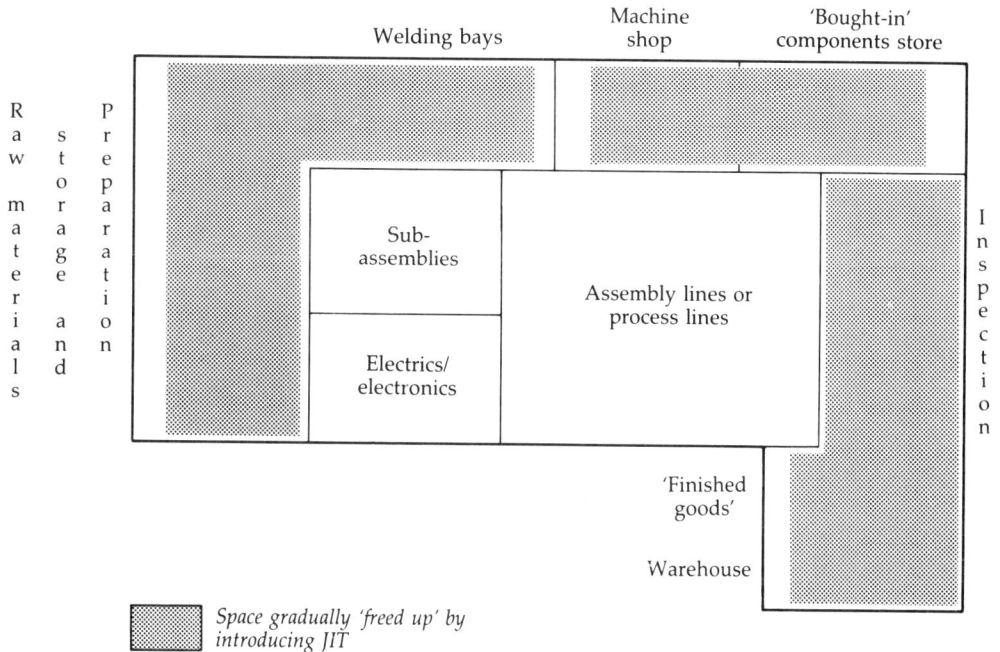

Raw materials · Storage and · Preparation

Sub-assemblies

Electrics/ electronics

Assembly lines or process lines

Inspection

'Finished goods'

Warehouse

Space gradually 'freed up' by introducing JIT

Figure 5.5 Subdivisions affected by JIT

needed for the small batches of the inevitable remaining B and C lines, as well as for any special preparation process which is not economical, or qualitatively practical to source externally.

2 Often whole welding bays are closed down in favour of a complete sub-contracting policy – again in many cases, obviating the need for fabricated component storage, as the welding sub-contractors increasingly provide a JIT service for the standard volume component runs.

3 A good portion of the machine shop may be kept on as reserve capacity for bespoke work – it may even offer an aggressive sub-contracting service of its own to external customers, hence maximizing machine utilization.

4 It is not uncommon for companies to adopt a policy of closing down the bought-in components store completely, in an assertive posture to ensure that what components are ordered slightly ahead of the job (e.g. up to half a day ahead) go straight to their appropriate work station as emergency buffer stock, to be stored on racks adjacent to the work station. Inputs are often ordered by the operative of the work station.

5 Like the machine shop, the 'electrical' bay will certainly survive, as the demands for more and more automation on whatever products are on offer, dictate the need for prototyping and testing bespoke computer control boards prior to sourcing them outside.

6 The other virtually redundant space is the finished goods warehouse – often the easiest and first to go completely as the company gains confidence in meeting same-day demands in a function over which it controls its own destiny. The level of finished goods is usually inversely proportional to the degree of confidence that the marketing department has in

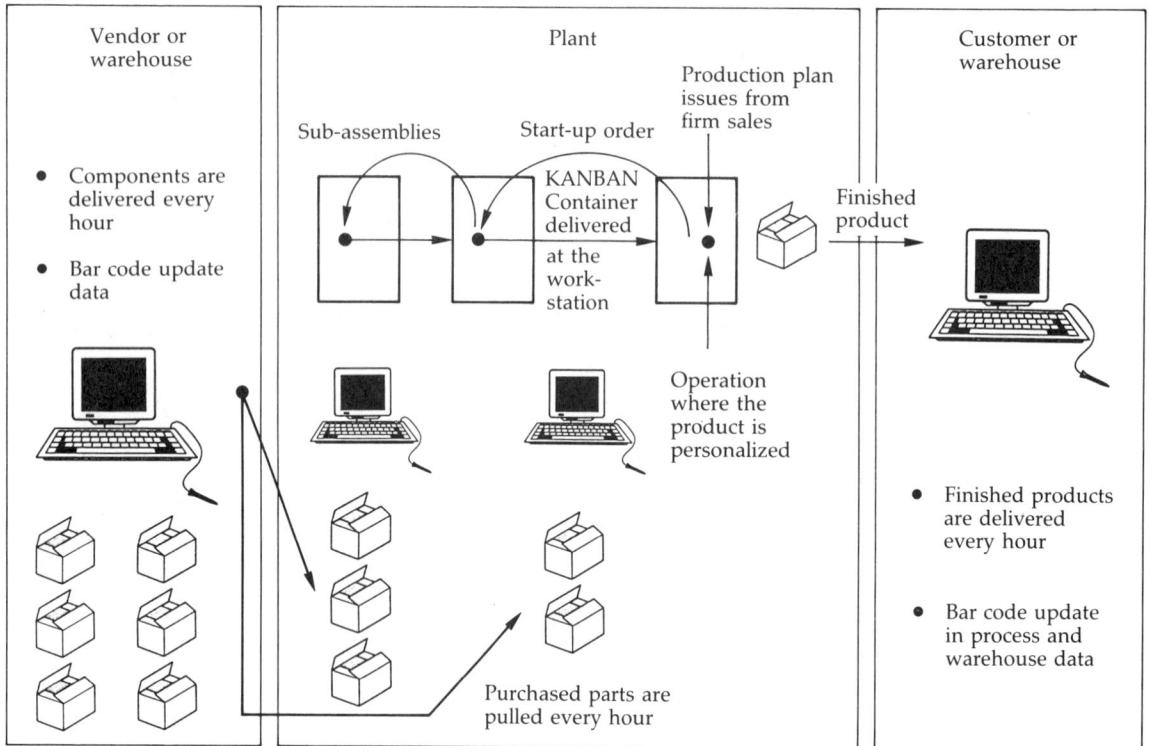

Figure 5.6 The JIT process

the manufacturing department! In other words, the higher the confidence, the lower the level of finished goods buffer stocks.

7 Although the outstanding space is still needed, the diagram belies the extent of reorganization which occurs here also – this is the heart of the Kanban system and FMS cells. The erstwhile sub-assembly area is still needed for the production of goods which fall outside the Pareto criteria for action.

The sophistication available in modern manufacturing technology is a subject in its own right, and beyond the scope of this book. It is a continuously evolving subject with developments occurring at bewildering speed, because this is a modern company's cornerstone for lasting competitive edge. The best way for students to keep up to date is to visit companies where state-of-the-art manufacturing technology has recently been installed. The surprise is that the IT needs of JIT are often simpler than the present generation of production support packages. They are often simple stand-alone islands of automation, linked together under CIM to generate Kanban triggers, restocking and statistics for estimating flow rates and quality measures to name a few. Overall then, the traditional assembly lines for high volume products (those identified as JIT targets by the Pareto criteria), give way to a series of integrated cells, inter-linked by various denominations of pull systems based on the Kanban principle, as shown in Figure 5.6.

The potential of modern manufacturing

Below are some lessons learned, and some impressive results already achieved by UK companies who have adopted the JIT philosophy as a way of working. Practice being a long way from theory, it is worth emphasizing that JIT just doesn't happen overnight. It requires a disciplined and targeted approach over extensive time periods – three years is not untypical, and, as discussed in Chapter 12, a permanent sea change in attitude – which can only be brought about by concerted retraining programmes.

In summary:

- Manufacturing processes are a combination of bespoke, batch and continuous processes in reality. Not all can be turned over to JIT.
- Because the changeover must presumably take place while the traditional business still operates, many companies opt for trying JIT out on a pilot basis.
- They often use a hand-picked project team to kick-start the process. It should be noted that benefits derive from a range of integrated actions which, when combined together, provide outstanding results. But it will not work without dedicated commitment to the philosophy once the pilot projects have proved their worth.
- Careful monitoring then follows, while the basic concepts are adapted to the unique needs of each company. Trade-off decisions are always required between what constitutes a 'volume line' justifying JIT treatment, and what are smaller batching processes: for instance for more bespoke sideline products, would they continue to operate if isolated from the new approach?
- JIT Kanban cells can in reality be introduced on 50 per cent plus of 'push' systems.
- Productivity increases of 30 to 40 per cent are possible.
- Stocks can be reduced by 50 to 60 per cent. Example: a £20 million business with £4 million stock has been able to release capital of £2 to £2.4 million, for reinvestment or reduction in borrowings.
- Floor space can be reduced by 30 per cent.
- Manufacturing lead times reduced from one week to one hour.
- Indirect functions can be reduced by over 50 per cent.

Conclusion

In this chapter we have given the briefest outline of how modern manufacturing systems operate. However, it must not be forgotten that the marketing of products designed and produced by such systems must also be planned at the earliest stage. Thus, product development and design need the widest variety of skills to integrate design and production into a marketing led strategy. This dictates an integrated teamwork approach, rather than function separation and the loose co-operation of specialists. Consequently, management philosophies and organizational practices are also undergoing a revolution, whereby new working methods are being forced on business by the marketplace. Project management, and its multi-disciplined nature is covered in the final chapter, while the implications of manufacturing methods for costs and strategies are picked up in Chapter 10.

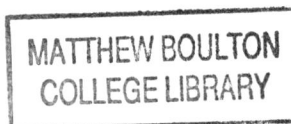

Chapter summary

In this chapter we have examined flexible manufacturing systems (FMS), design for manufacturing (DFM), customer activated production, by the Kanban system, materials control using JIT and quality management. It is argued that these are the major elements of a modern production system, and they are used partly to replace and partly to compliment more traditional production techniques.

Acknowledgement

Figures 5.1 and 5.2 are adapted from a conference paper (1986) by Professor J. Parnaby: *Cambridge Manufacturing Forum. Winning in 1990.*

Further reading

Many basic management textbooks offer little explanation of the elements of this chapter, although the issues are outlined in *Management Theory and Practice*, by G.H. Cole (DPP, 1990). The complexities of JIT and other elements of modern production systems are well explored in *Just-In-Time Manufacturing in Perspective*, by A. Harrison (Prentice Hall, 1992). *Total Quality Management*, Second Edition, by J. Oakland (Butterworth-Heinemann, 1993) has become the standard text on quality issues, although readers in a position to implement TQM might benefit from reading *Total Quality Management: Three Steps to Continuous Improvement*, by Tenner and DeToro (Addison Wesley, 1992) as well.

Tutorial exercises

5.1 It is argued that TQM is free. Why would this be so?

5.2 Consider the case of Notts Knitting in Chapter 2. In what way could the following methods improve the business, and what difficulties would arise if they were introduced:

(a) JIT

(b) Quality circles

(c) CAD/CAM

6 Financial accounting

In Chapter 1 it was stated that accountants and engineers traditionally had little in common other than professionalism. More recently, use of multi-disciplinary project teams, together with related changes in organizational structures and methods, have increased the interaction between the disciplines. This does not imply that engineers need to understand the detailed knowledge base of the accountant, but they do need some empathy with, and appreciation of, accounting to improve communications between the two groups, and reduce the misunderstanding and frustration that can arise (as would an appreciation of engineering by accountants).

It is the intent in this chapter to identify the traditional major outputs of the accounting process that are common to all organizations: profit-making and non-profit; private and public; large organization and small; and whole entities and subdivisions of those entities. In order to appreciate the traditional outputs of the accounting process some preamble is required to identify other aspects of the accounting and financial spectrum that will be discussed in subsequent chapters.

For the purpose of exposition the whole of the accounting subject matter will be divided into three traditionally broad categories:

1 Financial accounting.
2 Management accounting.
3 Financial management.

These categories can be broadly represented through a time-based model as illustrated in Figure 6.1.

The horizontal axis is a time dimension representing the past, present, short-term and long-term future of an organization's activities. The symbol 't' will represent time. Thus t_0 represents the present, t the short-term future, t_{-1} the recent past, t_n the long-term future and t_{-n} the long-term past. For the purpose of clarity we will assume that each t is a discrete time period of one year. In Phase 1, t_{-1} to t_0, it is the express purpose of financial accounting to measure the economic transactions undertaken by various members of the organization and report them in a summarized format to the external fund providers. In Phase 2, t_0 to t_1, it is the purpose of management accounting to provide routine and *ad hoc* reports to the regular economic activities undertaken by the organization, particularly with respect to cost control. Finally, in Phase 3, t_0 to t_n, it is the purpose of financial management to ensure that the organization's future investments will be economically viable and provide satisfactory returns to the owners. Prudent financial management also requires consideration of the organization's ability to raise external finance, and justify its use in improving the economic benefits to capital providers.

Figure 6.1 Time
horizons and levels of
accounting

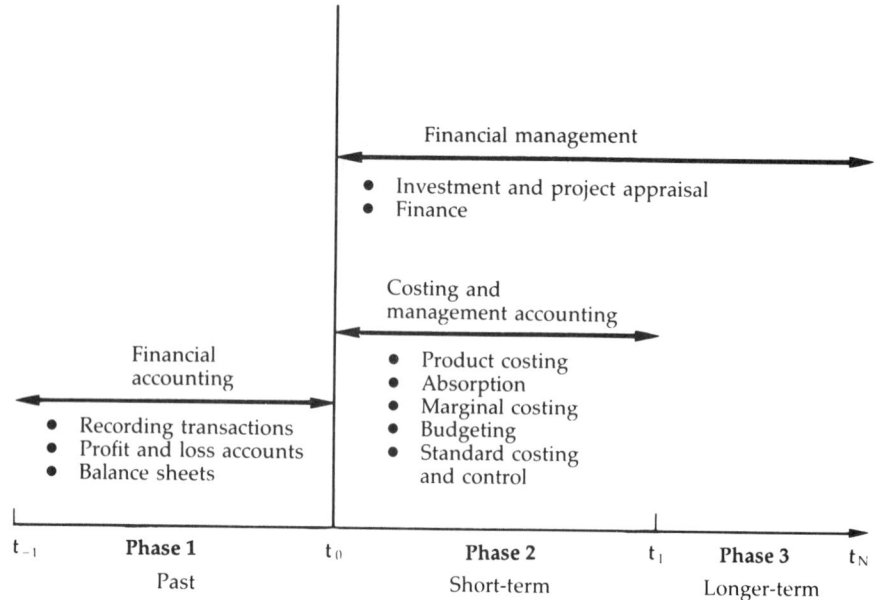

Figure 6.1 Time horizons and levels of accounting

As time passes, the model should reflect the change in accounting as it is carried out. Rather than this being a continuous process, as with engineering development projects, accounting is generally dealt with in discrete periods. For example, when t_1 elapses in one year's time, it will become the new t_0 (if the measuring process were capable of being continuous with advanced technology then tomorrow means that a new t_0 needs to be identified). Similarly, the original t_0 becomes the past, and when one year elapses becomes t_{-1}.

This model has significance when trying to identify the true nature and value of accounting information for the wide variety of user groups, of which the engineer will be but one subset.

Further, every organization may display its own version of the very generalized model as depicted in Figure 6.1. For example, a small civil engineering sub-contractor may possess a small financial accounting capability, but little else. Whereas a large construction company operating internationally may possess, at the very least, all of the accounting functions mentioned above, albeit with a greater degree of overlap and interdependence between the three areas. This chapter will concentrate on a brief explanation of some of the major inputs, processes and outputs of financial accounting. Greater emphasis however will be placed on the outputs of financial accounting, as it is the authors' view than an understanding of the information content of accounting reports and their inherent limitations will assist the engineer in appreciating the underlying causal factors behind what might be perceived as irrational or 'illogical' decisions and directives that contradict their own instinct and trained logic. For example, poor financial 'results' (measuring t_{-1} to t_0) may require a reappraisal of the marketing objectives of the firm which, in turn, might cause deferral or 'mothballing' of design and development projects, prototyping and, for example, delaying investment in the latest CAD systems.

The next chapter will examine the uses and interpretations that external users may make of financial accounting outputs.

Chapters 8 and 9 will examine the major management accounting inputs, processes and outputs. Chapter 11 will provide an introductory exposition to the nature of long-range financial management.

Forms of business entity

Organizations can take a wide variety of forms, from small-scale self-employed building sub-contractors, to major multinational civil engineering firms. These diverse business types are referred to collectively as business entities. In this section we briefly review each entity.

A construction engineer wishing to set up his or her own firm will need sufficient funding to begin with. If he or she wishes to be the sole owner, he or she will have to provide capital. This may come from several sources, either the owner will use savings (meaning that the firm then effectively owes money to its owner) or from external sources (meaning that the firm will owe money to a bank or other lender). Once the capital has been raised, the funds must be effectively managed in order to achieve the overall objectives of setting up in business in the first place – normally making a profit.

In this business, the proprietor (often called a sole trader) will generally be the manager. Nevertheless, from an accounting perspective, the provision of the funds should be separate from the use of the funds; hence the business is said to owe money to the owner. From a legal perspective the business capital and resources, and the owner's private belongings, property and possessions, are one and the same. If things go wrong for the business, the proprietor's personal assets (such as house, car, investments etc.) could be put at risk.

Similarly, partnerships involve more than one owner providing the capital, up to a specified maximum number of partners. The initial funds provided by the partners are considered as capital. Partnerships can be more effective than sole traders when the resources required to get started or grow are beyond the means of one person. In addition, the range and depth of management skills is often increased.

The management of the partnership is usually undertaken by the partners themselves, although managers can be appointed as in large civil engineering businesses. Like the sole trader, the partners are responsible and legally liable for the debts of the partnership from their own personal assets.

The entity common to large engineering businesses is the company, found in two main forms; the private limited company (Ltd) and the public limited (Plc). Both types of company are owned by shareholders, who may not be the people who originally set up the enterprise. The principal difference between the two is that ownership of shares in a private limited company is somewhat restricted by the existing shareholders, whereas any person, institution or business entity, can buy and sell shares in a public limited company. Consequently, there is a large secondary market in the shares of public limited companies. Shares of large companies are traded on the Stock Exchange, while those of smaller companies are traded on the Unlisted Securities Market and Third Market.

Ownership (and influence) in a company is represented by the possession of shares. Generally, the number of shares held in a company indicates the degree of ownership of that company. Few shares per individual would mean only a very small ownership in a Plc, whereas many shares not only makes ownership significant (and probably expressed as a percentage of the shares issued by the company) but also provides greater influence in the future direction of the business. Fifty-one per cent of share ownership would be enough to exert a great deal of influence over the remaining 49 per cent when the time comes for formal shareholders' meetings such as Annual General Meetings (AGMs).

There are two main types of share: ordinary and preference shares. The ordinary shareholder is entitled to a pro rata amount of the dividends awarded by the directors of the company each year which are, broadly, paid out of the net profits after tax generated by the company. The amount of dividends awarded will depend on a number of factors of which the amount of profit made is most significant, although future investment plans, the expectations of the investors and any future liabilities to other external parties will also play a part.

Dividends can vary up or down, year on year – or even not be paid at all. The shareholders can have no legal objection to a poor or non-existent dividend, although they can influence matters by voting at AGMs if not happy with the current directors' decisions.

The preference shareholder is similar in status to the ordinary shareholder except that when dividends are paid, this class of share is given priority over the ordinary shareholder. The dividend paid to the owner is generally fixed, and will normally be expressed as a fixed percentage of the normal value (see below) of the preference share. When profits are high, and high dividends are paid to ordinary shareholders, the preference shareholder will not receive dividends over and above the fixed percentage rate. Preference shareholders do not normally have a vote at an AGM.

Shares are expressed in nominal amounts such as 25 pence. If a company issues 2,000,000 shares at 25 pence, then the nominal value of the shares will be £500,000. Owning 1000 shares means ownership of 0.05 per cent of the company. Owning 1,000,000 shares means owning 50 per cent of the company. If a dividend of £100,000 was paid out of profits at some point in the future, then the holder of each share will be entitled to £100,000 / 2,000,000 shares = 5 pence per nominal value of the share. Owning 1,000,000 shares will therefore entitle the holder to dividends of £50,000. The dividend is the cash reward for investing in the company.

If the 2,000,000 shares were the only shares issued, and there was no form of loan, then the capital entrusted to the company will be 2,000,000 × 25p = £500,000. But if the company should wish to expand, it may need to issue further shares at some point in the future. Although the nominal value of the new issue of shares would still be 25p, the company may be able to obtain more from new and existing shareholders. If, for example, a further 1,000,000 shares were issued at £1.50 then, £1,500,000 will be raised (in addition to the existing £500,000 already raised). The additional 125 pence charged would be a share premium over and above the nominal value. Hence, 3,000,000 shares will be in existence. With a nominal value of £750,000, the share premium will be £1,250,000.

Any future dividend payment is expressed in proportion to the nominal value of the shares only. Any reserves, from past profits not distributed as dividends to shareholders, or from revaluing assets which have appreciated, also belong to the shareholders and are considered as part of shareholders' capital, or equity. There are specific rules laid down by a series of Companies Acts that give companies a separate legal status to that of their owners. Consequently, limited companies are distinct from other forms of business entities in two ways:

1 Because of the nature of the business the capital providers (the shareholders) are taking on a higher amount of risk with the sums they provide to the company. If the sums provided are managed well, then, their wealth will increase over time. The reverse could be the case, so much so that their original shareholding may be completely lost. But, unlike the other business forms discussed above, the shareholder will only lose the amount originally invested in the company, and personal property cannot be forfeited.
2 There is often a separation between the owners (shareholders) and the managers (controllers). In the large Plcs, this is very likely because the owners tend to be large institutional investors such as pension funds, insurance companies and investment trusts. Their detailed technical and engineering knowledge is likely to be less than the industry professional managers and directors they employ.

Engineers of all disciplines will be employed in other types of organizations where the overriding goals are not simply profit, such as local authorities, universities, health authorities and central government departments. These organizations are still subject to regulatory agencies and disclose information about use of the resources entrusted to management.

All of the above types of organizations, business and otherwise, will need to provide periodic financial statements, either because of statutory requirements in the case of companies and some public sector organizations, or in the case of sole traders and partnerships, so that returns can be made to the Inland Revenue. Preparing and evaluating these periodic statements is the work of the financial accountant.

Context of financial accounting

Engineering businesses and organizations will need to record systematically every measurable financial transaction. Transactions will include payments for the raw materials needed to manufacture a product, and receipts when the products are sold to the customer, plus payments for wages and salaries, energy costs, telephone bills etc. They will include purchasing technologically advanced manufacturing control systems, 4GL computers, robotics from suppliers on credit and payment for leases of equipment. Transactions are recorded in a standardized manner, which has not changed substantially for many centuries, commonly known as double entry bookkeeping. It is important to bear in mind that most of the accounting principles and techniques discussed in this book will depend crucially on accurate and systematic bookkeeping. A brief explanation of the double entry system is appended to this chapter.

Organizations, business and otherwise, are expected to comply with a range of specific statutory obligations to describe the resources entrusted to the managers (or appointed officials) and how those resources have been used. These resources are expressed in financial terms, and structured to satisfy the following equation:

Liabilities = Assets

The equation forms the key to a periodic summarized financial statement called the balance sheet. The balance sheet is simply a representation of the equation above at measured periodic intervals, intended to provide a concise overview of the organization's financial condition at a particular point in time.

The terms liabilities and assets are part of accounting terminology. Another way of expressing the same equation would be in terms of measurable financial resources:

Resources owed Resources owned
by the organization = by the organization

(or funds owed by the organization to various external parties = funds converted and managed by the organization's agents, namely, engineers and managers).

In its broadest context, liabilities are funds owed to parties outside the organization. There may be a large number of these, including: capital providers (e.g. shareholders), lenders (e.g. banks), and creditors (e.g. suppliers of materials on credit, Inland Revenue, etc.).

Assets are the conversion of these liabilities into identifiable items that the organization needs in order to fulfil its particular business objectives. These items will include: materials, machines, computers and vehicles. The liabilities (funds owed to external parties) can be regarded as a pool of funds that have been transformed into assets by engineers and managers in order that the business can conduct its activities and achieve its goals.

The balance sheet

Because the balance sheet is representative of the accounting equation: assets = liabilities, emphasis will be given to the basic principles of balance sheet compilation. Using a series of simple financial transactions the principles of the balance sheet will be demonstrated.

Example: The N. Gineers

A balance sheet after the first day of a business's existence could be summarized through one key financial transaction: the owners, Mr and Mrs N. Gineer, deposit their saving of £100,000 into a special business account at their bank.

Balance sheet as at Day 1 19×2

	£		£
Capital*	100,000	Business bank account	100,000
	100,000		100,000

*This capital might have been obtained in a number of ways, depending on the type of business entity used. In a company, the capital might comprise 100,000 £1 shares spread over the Gineer family and friends.

The accounting equation Asset = Liabilities still holds, since the business is now liable for £100,000 of owners' capital, but has an asset of £100,000 in the bank.

During the week, the funds were used to purchase plant and machinery and buy building materials. The plant and machinery cost £60,000 and the building materials purchased from a supplier cost £20,000. The effects of these transactions are shown on the balance sheet below:

Balance sheet for N. Gineer as at Day 7

	£		£
Capital	100,000	Plant and machinery	60,000
		Stock of building materials	20,000
		Business bank account	20,000
	100,000		100,000

The composition of the assets has now been changed with two new transactions: from cash in a bank to the physical ownership of machines and raw materials. The documentation involved with the transactions would be important supportive evidence of these transactions, namely, a receipt would be provided by the supplying company in each case. The capital funding of £100,000 hasn't altered. Assets still equal liabilities.

Tomorrow, day 14, and subsequently, wages and salaries will have to be paid and some of the raw building materials used. These activities will continue until the construction is complete and the customer takes possession in exchange for the agreed contract sum. This will necessitate a series of transactions that will need to be recorded systematically.

In the second week (days 8 to 14) £15,000 of building materials are used by skilled labourers, who were paid £12,000. Again, it is the composition of the assets that change, because of the nature of the transactions to date. The nature of the raw material has changed because some of it is being used in the construction of the final output. This amount needs to be reflected in a different category of asset called work in progress (WIP). It is quite likely that further resources will need to be used on this before the product is in its final state and ready for sale. Until this final state is reached, the WIP figure will increase continually, to reflect the resources used to date. The creation of the WIP asset is accomplished by the expenditure of resources available to the business in the form of wages paid to labour (in this case £12,000 paid to craft workers and £15,000 paid for materials).

Balance sheet for N. Gineers, as at Day 14

	£		£
Capital	100,000	Plant and machinery	60,000
		Raw materials	5,000
		Work in progress	27,000
		Bank balance	8,000
	100,000		100,000

Consider the previous balance sheet at day 7 and compare with the balance sheet at day 14 after accommodating the transactions above. The balance sheet will be presented in the form of:

Assets = Liabilities

This vertical layout is the usual format for published accounts and will be used from here on.

Extract of balance sheet as at Day 14

Assets	Day 7 £	Transaction	Day 14 £
Plant and machinery	60,000	No change	60,000
Stock of raw materials	20,000	Used £15,000	5,000
WIP stock	–	£15,000 of raw materials processed by labour of £12,000	27,000
Bank	20,000	Paid £12,000 in wages	8,000
	100,000		100,000
Liabilities			
Capital	100,000	No change	100,000
	100,000		100,000

The fundamental equation:

Liabilities = Assets

has held true throughout. The composition of the assets has changed since Day 1, but the liabilities, in this case to the owners, still consist of the original £100,000 capital only.

Activity

In the N. Gineer example above, further manufacturing and engineering activity was needed to complete the product in a state readily saleable to a customer. By day 21, this had required the use of a further £3,000 raw materials and a further £2,000 to be paid in wages. Produce the full balance sheet as at day 21 and show workings.

Answer

Balance sheet as at Day 21

Assets	Day 14 £	Transaction	Day 21 £
Plant and machinery	60,000	None	60,000
Stock of raw materials	5,000	Used £3,000	2,000
WIP stock	27,000	Processed £3,000 of raw materials with £2,000 of labour	

Bank	8,000	Paid £2,000 in wages to WIP	32,000 6,000
	100,000		100,000
Liabilities			
Capital	100,000		100,000
	100,000		100,000

Once the products are finished and ready for resale, a 'dummy' transaction takes place whereby the stock of WIP becomes a stock of finished goods.

On day 24, sales and marketing activity enabled the completed product to be sold to the customer for cash. It would be appropriate to try to obtain a price to the customer that is greater than the cost of buying and manufacturing (though for tactical marketing reasons this is not always the case in the short-term).

Total cost	£32,000
+ Mark-up 25% × 32,000	£8,000
= Price	£40,000

Although there is essentially one transaction, there are two major effects on the balance sheet:

1 The finished stock to be transferred to the customer is removed from the balance sheet and is replaced by cash. This would simply be a transfer of one item of the balance sheet, finished stock, into another, cash, if, and only if, price was the same as cost. But:
2 Because price, in this case, is higher than cost, there is another effect: the wealth of the owner's capital increases by the difference between cost and price, i.e. Profit.

The balance sheet using the assets = liabilities in vertical format on day 24 should read thus:

Balance sheet as at Day 24
£

Plant and machinery	60,000
Material stock	2,000
Bank	6,000
Cash	40,000
	108,000
Capital	100,000
plus profit on day 21	8,000
	108,000

In reality, it would be unusual for such a large bill (i.e. £40,000) to be paid in cash. If the customer had paid with a cheque, the bank balance figure would have increased instead. However, a great many sales are made on credit. Since no cash is received for a credit sale, the balance sheet is adjusted for the sale by removing the finished goods stock and adding profit in the way described above, and the debt recorded as an asset (referred to as trade debtors, or simply, debtors). At some future point the debtor will pay all or part of the promised price. This payment will be reflected by a decrease in the debtor's figure, and a corresponding increase in the cash or bank balance figure. Similarly, the company may purchase raw materials on credit, rather than pay cash or write a cheque. In such a case, the bank balance or cash figure would be unaffected by the transaction, but until the debt is paid off, the business will have a liability. This liability is referred to as trade creditors (or just creditors) and recorded as a current liability.

Activity

1 Consider the effects on the balance sheet of the following:

    a. the product sold for only £30,000 on day 24 instead of £40,000?
    b. the £30,000 is not due to be paid until two month's time

2 To what extent does the Gineer's decision to price at cost plus 25 per cent reflect good marketing practice?

Long term and short term

In any given period, the building company will wish to repeat the above activity as frequently as possible, within its resource constraints. Therefore, a number of these transactions will be taking place simultaneously and systematically throughout the business.

Some types of transactions will be very repetitive in any one period, while others (such as capital) will not. The balance sheet, in practice, tends to reflect this in its presentation. Buying plant and machinery are not as repetitive as the more routine transactions of buying materials and paying wages. Thus further categories are introduced into the balance sheet to show long- and short-term assets and liabilities separately. The balance sheet equation is modified to reflect these categories:

i.e. Liabilities = Assets becomes:

Long term liabilities   Fixed assets
+   =   +
Current liabilities   Current assets

On the liabilities side of the equation where long term liabilities (from now on referred to as capital employed) are funds provided to the organization for use over a succession of time periods as opposed to current liabilities, normally regarded as debts which are usually to be paid within one year. Assets purchased in order to benefit the organization over future discrete time periods (t_1, t_2 to t_n) are regarded as fixed. Assets likely to be used up (and replaced) frequently within one discrete time period are considered as current assets. So a company car used by the chief engineer is a fixed asset for the business, but a similar car purchased by a motor trader for resale would be classed as stock; a current asset.

The original equation still holds, but the above variation is simply a reflection of sub-sets of each category. Furthermore, there are additional sub-sets.

On the liabilities side:

Capital = Shareholder funds + long term loans

Current liabilities = Creditors + bank overdrafts + Taxes owing
_____

= Total liabilities
_____

On the assets side:

Fixed assets = Land and buildings + Fixtures + Machinery + Vehicles
+

Current assets = Stock + Debtors + Cash in bank + Cash in hand
_____

= Total assets
_____

(It is worth noting that the categories in the balance sheet are beginning to reflect the interests in the business of external parties: capital providers, banks, creditors and the Inland Revenue etc.).

Before moving on to another major financial statement – the Profit and Loss account – it is worthwhile emphasizing the logic of recording transactions once more.

The transformation process for manufacturing engineering goods can be broadly summarized as:

Raw material $\rightarrow$ WIP $\rightarrow$ finished stock $\rightarrow$ sale on credit $\rightarrow$ cash

This transformation process generates a series of financial transactions resulting in a series of balance sheets, where the accounting equation always holds.

Example: Sam's Sewing Aids

After his retirement from Notts Knitting, Sam set up Sam's Sewing Aids, making specialist components for his former employer.

Consider the following simplified opening balance sheet where, four weeks ago (t_{-4}), the business began with £5,000 capital and the cash provided was used to purchase stock for £3,000 and the remainder of £2,000 was banked.

		£		£
Balance sheet t_{-4}:	Capital	5,000	Material stocks	3,000
			Bank	2,000
		5,000		5,000

Three weeks ago, £1,000 of the bank account was used in the payment of wages. The material stock used, together with the wages, becomes an incomplete stock (called work in process or WIP) that costs £4,000 (i.e. £3,000 materials + £1,000 wages).

		£		£
Balance sheet t_{-3}:	Capital	5,000	WIP	4,000
			Bank'	1,000
		5,000		5,000

Two weeks ago, the WIP stock was completed with further labour activity of £1,000 and the completed product was physically transferred to the warehouse.

		£		£
Balance sheet t_{-2}:	Capital	5,000	Finished stock	5,000
		5,000		5,000

One week ago, the finished stock was transferred to the customer together with an invoice for £8,000.

		£		£
Balance sheet t_{-1}:	Capital	5,000	Debtor	8,000
	Profit	3,000		
		8,000		8,000

Today, a cheque was received for £8,000 from the customer. The obligation of the customer to the firm has now been discharged.

Note, the profit had already been measured in the previous balance sheet t_{-1} but the cash was received in the following balance sheets.

		£		£
Balance sheet t_0:	Capital	5,000	Bank	8,000
	Profit	3,000		
		8,000		8,000

As stated earlier, the provision of credit is usually a two-way process, with the firm buying materials on credit as well as selling on credit. The acquisition of materials on credit creates a current liability of creditors on the balance sheet. For example, Sam wished to continue trading by purchasing £4,000 of materials from a supplier on credit in period t_0.

		£		£
Balance sheet t_0:	Capital	8,000	Cash	8,000
	Creditor	4,000	Raw materials	4,000
		12,000		12,000

The profit and loss account

The Profit and Loss Account (PLA) is the major accompanying statement to the balance sheet. Engineering activity tends to have a great impact on this account. Unlike the balance sheet, the PLA is a statement that aggregates the costs of all the resources involved in producing and selling products over a period of time, and deducts these from the sales revenue earned from selling those same products. The term expenses is often used interchangeably with costs as the measure for periodic resource usage.

Sales revenue is, at a minimum, a function of two variables: the price charged to the customer per item and the quantity of items sold. One project completed between t_0 and t_1 at £500,000 will mean sales revenue of £500,000 for the firm's PLA. Five projects valued at £500,000, £750,000, £350,000, £1,200,000 and £150,000 respectively started and completed between t_0 and t_1 will generate revenue to the firm of £2,950,000.

The role of the PLA is to match the expenses and revenues earned by a firm over a period of time and determine the profit (or loss) for the same period:

If Sales > Expenses = Profit
but, if Sales < Expenses = Loss

In the first case the resources used in producing the product have generated a return from customers that is greater than the use of resources. The surplus, profit, is the reward to the owners of the business. The result will be an increase in the owners' wealth. In the latter case, the reverse is true, and a deficit or loss has resulted. The owners' wealth has therefore decreased.

As a starting point, the N. Gineers' case looked at earlier in the chapter could be shown thus:

Profit and loss account for the period ended day 24

	£	£
Sales less:		40,000
Cost of goods sold	32,000	
Gross profit		8,000

The N. Gineer case is highly simplified for two reasons. First, the process of converting raw materials into a finished saleable product is rarely so straightforward that skilled labour is the only input required. The workshop would normally create other expenses, such as energy costs, supervisory labour etc. These expenses, which are incurred in the production process itself, are shown in the PLA.

The full layout of the PLA showing manufacturing expenses is developed in Chapter 8 on page 175.

The second over-simplification of the case is that there are many other expenses not directly associated with the production process itself. These would include salaries of sales, administrative and marketing personnel, advertising, travel, and interest payments, among many others. These expenses are often called revenue expenses, since they are incurred in achieving the revenue for the current measurement period under consideration, but are not costs incurred by producing the product directly. Thus the PLA for N. Gineers might look so:

Profit and loss account for the period ended day 24

	£	£
Sales less:	40,000	
Cost of goods sold	32,000	
Gross profit less:		8,000
telephone	100	
heating and lighting	150	
postage	80	
salaries	1,500	
advertising	275	
travel	93	
		2,198
Net profit		5,802

The final net profit figure will be transferred to the balance sheet as an increase in the Gineers' capital, and not the £8,000 figure shown in our simplified example. It is important to recognize that the accounting equation, Assets = Liabilities, will still hold. In this case, N. Gineer paid these expenses by cheque, thereby reducing the bank figure by £2,198 – but the profit is also low.

The relationship between PLA and the balance sheet is important in the accounting process:

1 Normal business would suggest that the most repetitive transactions are buying and selling. Buying involves not only materials, components and finished product, but also wages, salaries, electricity, water etc.
2 If these transactions could be dealt with by segregating the profit and loss items from the balance sheet, the task of providing a new balance sheet is avoided every time a transaction takes place.
3 The reason for this is that only one profit figure is determined for the period of consideration, normally one year, and this total profit figure will be added to the capital figure in the balance sheet one year later.
4 The tally of day-to-day transactions is kept in the form of revenues and expenses and totalled at the end of the period.

The profit and loss account will reflect the trading and business activity over the accounting period expressed in financial terms and will measure the profit or loss of that same period.

A balance sheet will identify, at the point in time it was produced, the resources entrusted by the owners and other external funders, such as creditors, and will show that the liabilities will equal the total assets owned by the organization.

Its value as an informative document to outside readers can be greatly enhanced by a previous balance sheet – which would probably be one year ago. The two balance sheets could be compared to identify significant changes and altered relationships between, say, creditors and debtors, profits and capital used etc. Many of these significant relationships will be discussed in Chapter 7.

Activity

Establish the future balance sheet of N. Gineer if the materials were converted into finished goods along with the further use of labour costing £3,000 in cash, and sold for cash to a customer for £10,000? What would be the total liabilities after the suppliers have been repaid?

Measuring accounting profits

This section is an important refinement of the measure of profit stated earlier. In arriving at accounting measures of profit there are some fundamental adjustments that have to be made at the end of the accounting period. These adjustments embody the concepts and conventions of accounting stated at the end of the chapter. The following major adjustments will need to be made before accounting profit for a period can be measured. There are three aspects that will be considered for illustration purposes before a PLA and balance sheet can be arrived at. First, the issue of stocks, second, the issue of depreciating fixed assets and third, the notion of accruals and prepayments need to be taken into account.

1 Accounting and stocks

In order to understand this refinement a brief discussion of the physical nature of stocks is required. Once the need for stocks in many businesses is established then the accounting treatment of stocks will be explained.

Physical nature of stocks

Stock is often regarded as a physical buffer between the uncertainty of sales demand at any time and the more easily programmable production schedules. Thus no retail outlet would be able to stay in business without having sufficient stocks available as and when uncertain demand dictates.

However, there are other forms of stock in business organizations that need to be available to ensure a production flow without bottlenecks. Work-in-progress (WIP) and raw material stock are held largely to act as a buffer for such production problems by ensuring that items at an appropriate stage of production can be processed even if part of the production line is inoperative, or output is restricted in some way. The use of just-in-time (JIT) techniques can do much to reduce stock level, but only in exceptional circumstances is the need for stocks completely eliminated (see Chapter 5).

In a manufacturing organization three forms of stock are prevalent:

Raw material stores → WIP → Finished goods

All forms of stock are items of value that the firm owns and should at any point in time be regarded as assets. From a business point of view, the fewer stocks needed the fewer resources need be tied up in current assets.

From a balance sheet perspective, total stock values have to be shown for the external user. Internally, greater recording detail has to be provided in order to track and monitor the acquisition, distribution and use of stocks and loss of stocks through wastage, loss, pilferage, evaporation etc. Effectively, it means that the asset value is reduced and as a consequence so must capital employed and shareholder wealth be reduced accordingly.

Activity

Consider the nature of stocks in the following organizations: a local builder; a nationwide construction company; a car component manufacturer; a machine tools firm introducing Computer Aided Manufacturing (CAM).

Stock control

One of the first major computer applications for many organizations was to control stocks. Many organizations now operate computerized stock control systems of varying levels of sophistication. In the retail sector, such organizations as Laura Ashley using Electronic Point of Sale (EPOS) transactions at the check-out can immediately register the change in stock level at the outlet and trigger off re-order levels of particular items. (Even so it is still necessary for accounting to undertake a stewardship function on behalf of the capital providers at least once a year and physically check the quantity and value of all stocks.) If there is any discrepancy, then the lower stock value is likely to be recorded in the profit and loss, and profit reduced accordingly.

However, the final measurement of stocks at the end of a period need to be verified by accounting staff. This often requires a physical stock count. Sampling techniques are often used where a full check is not physically or economically appropriate.

The financial nature of stocks

The purpose of producing finished goods is to sell them, hopefully at a profit. While the stock remains with the firm it is treated as an asset and valued on the basis of the cost in acquiring it or converting it into a finished product. But, when the asset is physically sold (or title to the good is transferred to the buyer), the good (current asset) must be expensed against the revenue generated for the sale of that item.

This approach is fundamental to accounting and is known as the matching principle. Because, at the balance sheet date, there is likely to be stock remaining a systematic approach is taken in measuring total profit for the period:

Stock used = Opening stock (t_{-1}) plus production (t_{-1} to t_0) less
Closing stock (t_0)

The above equation when expressed financially is known as the cost of sales or cost of goods sold (cogs). It is the cost of sales that is matched against revenue to determine profit. In other words, stock left over from the previous period is added together with the production output made in the current period. Any stock remaining is treated as a current asset and sent to the balance sheet ready to be sold in the next accounting period. The matching principle is critical to understanding how accountants recognize profit. Hence, the profit and loss account reported externally invariably states this principle implicitly in its presentation:

Sales		X
Less cost of goods sold:		
Opening stock	X	
Add production	X	
Less closing stock	(X)	
Gross profit		X

Example

At the end of last year, a distributor of high tensile cable had 100 spools in stock, purchased at £200 each. During the present year, 600 further spools were purchased, also at £200 each. The distributor sold 500 spools during the year, at a price of £300 each.

The matching principle (see Appendix 6.1 on accounting conventions) would suggest the following:

	£	£
Sales		150,000
less cogs		
Opening stock	20,000	
Purchases for resale	120,000	
Less closing stock	(40,000)	
		100,000
Gross profit		50,000

Activity

Stock for Buster Bitt & Co, a car component manufacturer, remaining in the balance sheet from last year consisted of 200 completed carburettors for a customer that cost £8,000 in total to produce. This year 3000 were produced at a total production cost of £135,000. The sales for the period were 2700 at a selling price of £60 each.

Required

Calculate the gross profit for the end of the accounting period and value the remaining stock for balance sheet purposes.

The above example raises further accounting issues about the valuation of stock that will be addressed in Chapter 8. It should be evident from the question that the costs of producing carburettors have increased from last year. The realities of business life, particularly under inflationary conditions, are that the purchase of materials increases (as can wages and overheads such as energy and rates).

For external financial reporting purposes the method to be adopted here is known as first-in-first-out (FIFO). This implies that those components purchased at the older (lower) price are used first and the stock is valued accordingly. There are other methods, such as Last-in-First-out (LIFO) where it is assumed that the newer, more expensive components are used up first. These methods of stock valuation are discussed in Chapter 8.

If stocks did not exist, then the measurement of accounting profit would be much simpler, as in the case of design consultancy firms, who offer a service based on their expertise. In manufacturing firms, stocks are a reality and thus profit has to reflect the stock position by adjusting the reported value of raw materials, WIP and finished goods (or a combination of all three), to reflect increases and decreases in stock levels between the beginning and end of each accounting period.

Although stock levels which increase each accounting period give the impression of growing assets, caution needs to be exercised because it may suggest that production is dislocated from the level of sales and that unnecessarily large amounts of capital are being tied up in assets that may not provide any immediate or future return if not converted into final product and sold.

At the time of writing a number of firms were having to 'write-off' unsaleable stocks because of obsolescence, poor quality or a change in buying behaviour. However, if the general market prices for the goods are something less than the costs stated in the balance sheet, then accountants should exercise (on behalf of the external providers of funds) another important convention called 'conservatism' and reduce the value of the stock to the outside market price less any costs that would be incurred in disposing of the asset (which is referred to as the net realizable value (NRV) of the stock). The effect on profits of a large scale NRV or, in the case of no value, at all, to 'Write-Off' would be to reduce the reported profit in that period (as would any large scale loss of an asset due to theft, wastage, obsolescence, fraud etc.).

2 Accounting and depreciation

From a balance sheet perspective, capital investment in new machinery, plant and equipment is shown as a fixed asset. As stated earlier in the chapter, the difference between fixed and current assets is more to do with time than the nature of the asset itself: the former are intended to remain on the balance sheet beyond one year and benefit the organization over future periods; the latter should change in the very short term within the accounting period. Without the fixed asset, the basis for the organization to generate future benefits would not occur. Acquiring the fixed asset involves a cost. But the cost has been incurred to benefit a succession of future accounting periods, not just the current one. Clearly, the cost of the investment must be spread over all the years in which the asset will generate benefits to the firm, and not simply treated as an expense in the present accounts.

The distinction between current revenue expenditure and capital expenditure is important because this will determine the way the cost is treated in the final accounts. If the whole cost were incurred for the resource to be used during the current period only, then it would be written off in the profit and loss account and increase costs accordingly (thus, there would be no record of it being an asset in the balance sheet). If the business organization were making profits then they would be reduced by this cost. The higher the cost the greater the impact on profits.

If the asset purchased is considered as benefiting future periods, the cost of the fixed asset would need to be spread over the anticipated life of the asset in such a way as to reflect its use and diminution of original cost because of usage, obsolescence, wearing out etc. The sum deducted from the original cost is referred to as depreciation. The original cost of the asset is recorded in the balance sheet, together with the sum of the depreciation to date. It is only the annual depreciation of an asset that can be correctly called an expense. Depreciation is that proportion of the original cost of the fixed asset that will be charged to the profit and loss in each period. Thus the effect on each successive period's profits would not be as dramatic as a once-and-for-all charge during the period in which the asset was bought.

This annual charge could be even (meaning the same amount is charged each year) or uneven, depending on which one of several methods to calculate the annual charge is used. In the UK the most commonly used methods are:

A The straight-line;
B The reducing balance.

Straight-line depreciation

The straight-line approach provides for an even spread of the original cost.

$$\text{Depreciation (£p.a.)} = \frac{\text{Original cost of fixed asset}}{\text{Estimated life of the asset (years)}}$$

The numerator can be modified if it is reasonable to expect a commercial resale or salvage value at the end of the expected useful life of the asset. Thus depreciation becomes:

$$\text{Depreciation (£p.a.)} = \frac{\text{Original cost} - \text{Established residual value}}{\text{Estimated life of the asset (years)}}$$

Compared to the previous definition, the revised formula has the effect of reducing depreciation and thereby improving the profit figure. Of course, the higher the residual value at the end of the useful life of the asset, the lower the amount of total depreciation to be charged in the intervening years.

For example, a main-frame computer purchased by a firm of civil engineers cost £1.2 million. It was estimated that the useful life was ten years. Then the depreciation to be charged using the original formula would be:

$$\frac{1.2 \text{ million}}{10 \text{ years}} = £120,000 \text{ depreciation charge to PLA for each of the ten years}$$

More realistically, it might be considered prudent to reduce the estimated life because of the high obsolescence factor with computers to, let's say, five years, and since a second-hand user market exists, the company believes it could sell the asset for £120,000 in five years time. The revised calculation would be:

$$\frac{£1.2 \text{ million} - £120,000}{5 \text{ years}} = £216,000 \text{ depreciation per annum.}$$

The effects are a reduction in profits measured due to the depreciation charge. In fact where assets are likely to suffer from obsolescence due to technology advances, the conservative principle (see Appendix 6.1) takes over and the asset is 'written-off' to the profit and loss accounts as quickly as possible (thereby reducing profits further).

Reducing balance depreciation

The effect of the reducing balance approach is to charge the original cost at a declining rate over the life of the asset with the highest proportion of the cost charged in year one and, at a diminishing rate thereafter. This is sometimes known as accelerated depreciation where the amounts to be written off are higher in the earlier part of the asset's life. A more appropriate justification would be to consider the asset's income earning potential as diminishing in future years.

In some respects this approach is more realistic than the straight line method, since many assets, such as motor-cars and computer hardware and software tend to lose much of their value in the year of purchase. The mathematics of the method would suggest that depreciation be charged at a diminishing rate in perpetuity. This is unlikely to happen as the asset will have a finite life. If the life of the asset is estimated and the residual value is also estimated then the following formula will indicate the percentage rate at which to reduce the previous year's written-down value.

The reducing balance formula:

$$1 - \text{useful life} \sqrt{\frac{\text{Residual value}}{\text{original cost}}}$$

The above formula provides the percentage to depreciate each year from the written-down value. This would indicate the amount of depreciation for the year based on the previous period's written-down value.

For example, a heating-system was bought for £400,000 and was expected to last for ten years. The firm was expected to depreciate at 40 per cent per annum. What is the depreciation for the first three years?

	£	
Year 0 Purchase price	400,000	
Year 1 depreciation is 40 per cent of 400,000	160,000	– to year 1 PLA

Year 1 Net book value	240,000	– to balance sheet
Year 2 depreciation is 40 per cent of 240,000	96,000	– to year 2 PLA
Year 2 Net book value	144,000	– to balance sheet
Year 3 depreciation is 40 per cent of 144,000	57,600	– to year 3 PLA
Year 3 Net book value	86,400	– etc.

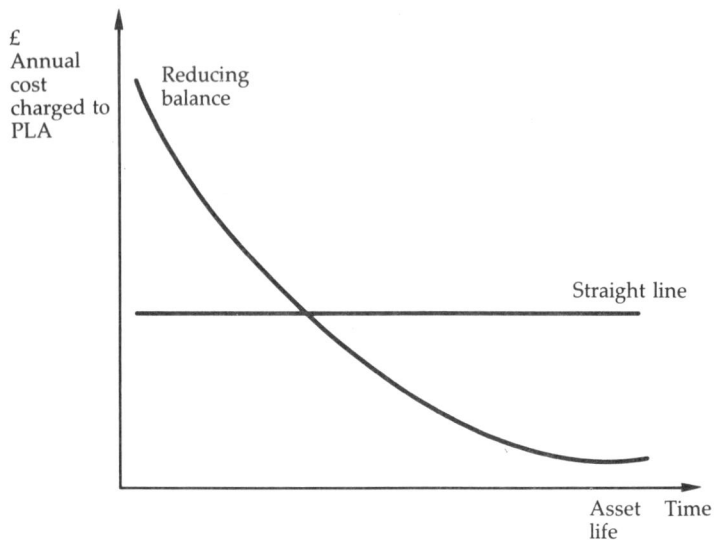

Figure 6.2 Effects of depreciation method used

The balance sheet will need to reflect the incidence of depreciation on its assets year on year. Normally, the assets are recorded in the balance sheet at cost. However, depreciation reduces the cost of the asset to its net book value each year. As the asset is likely to transcend a number of accounting periods, then the depreciation will need to be accumulated:

Extract of balance sheet
Assets at cost   X
less depreciation to date   X

Net book value   X

When considering the implications of two different methods using the same asset it is apparent that the reducing balance method will charge more depreciation in the earlier years using the above example than the straight line, whereas the reverse is the case in the latter years. This choice of method used will have implications for the profits reported over the life of the asset. Figure 6.2 shows the effects of the two methods cited on annual depreciation charges and the effects on reported profits over the life of the same asset.

If the asset were to be written-off before the end of its useful life then:

Original cost less depreciation to date = Amount to be written-off as a further expense in the profit and loss.

If the asset were sold before the end of its life or for salvage then:

Sale price less
(original cost less depreciation to date) = Amount to be charged to the
profit and loss.

This amount could be a profit on disposal or a loss on disposal:

If original cost – Depreciation = Net book value
(NBV) then where:

NBV > Sale price of asset = Loss on disposal;

and where:

NBV < sale price of asset = Profit on disposal.

Both profits and losses on disposal would be shown in the PLA.

It is worth pointing out that the reducing balance more closely resembles the UK tax position when calculating what organizations are allowed to offset against profits to reduce their tax bill. The Inland Revenue will use their own Treasury determined formula in calculating the tax obligation of the company. Thus depreciation would be substituted for 'Capital Allowances'. The system follows a reducing balance approach.

Depreciation is a real enough cost that needs to be reflected in each period's profit and loss whilst the fixed asset (from where the amount calculated arose) is reflected in the balance sheet. However, even though depreciation is clearly a cost to the business, no cash is actually paid to anyone. The distinction between cash costs, and non-cash costs, is fundamental to accounting and often causes misunderstanding unless the accruals concept (of which depreciation is an example) is understood.

Cash costs and non-cash costs

If the asset was bought for cash in the first place then the cash effect of buying the asset took place during that accounting period and only that period. Profits though would not be affected in that same period. However, over subsequent years, profits will be affected by the non-cash cost, depreciation, whereas the cash position relating to the asset would not be affected.

3 Accounting and accruals

If gross profit is the sales receipts less the cost of goods sold, then the calculation of net profit will require further deductions of cost. Most costs (or use of resource) will be paid for in cash during the accounting year that the resource was used, but evidence of cash payment is not sufficient to include it as a cost. Equally, cash receipts from a sale made in the last accounting year may be received in the subsequent year.

Accruals are costs which are recognized as an expense in a period before cash is paid out.

For example, the accounting year for a small manufacturing workshop is from 1 January to 31 December. It has just vacated its old premises and moved to a new trading estate. The let began on 1 November. There is a commitment to pay rent totalling £2,400 per annum with two six-monthly instalments, the first payable on occupation. Thus, on 1 November, £1,200 was paid to the landlord. However, for profit measurement purposes, only two months worth of the resource paid for was actually used within the present accounting period. The charge to the PLA will be two-sixths of the £1,200 i.e. £400. The remaining £800 is a prepayment for a service not yet received, and will be treated as an asset in the balance sheet.

If in the following accounting year, rent of £1,200 was paid in April and an administrative error overlooked payment in November, the charge for the year in the PLA will be £2,400. This will comprise of three elements:

1 The £800 prepayment becomes an expense for the period 1 January–30 April;
2 The £1,200 cash payment covering the period of 1 May to 31 October will also be expended;
3 £400 will be owing and is, therefore, an accrual that must be charged to the PLA.

The £400 accrual will need to be shown in the balance sheet as a current liability at the end of the accounting period.

Summary

All business organizations have to provide a summary of their financial transactions at least annually. The conventions, governmental and accounting legislative framework specifying the format and frequency of reporting will depend upon the type of organization companies have to conform to for the requirements of a series of Companies Acts.

Financial statements, as represented by the balance sheet and profit and loss account in this chapter, are periodic summaries of all the measurable financial transactions of a business.

The financial accounting process summarized

- Financial transactions are undertaken by the business.
- Source documents are evidence of the transactions.
- The essential details of the documents are transcribed into the firm's own listing format.
- The double-entry process is followed using ledgers or computers.
- At the end of a period, a trial balance is created ensuring that the above step has been properly recorded.
- Adjustments are made to reflect the measuring of stocks, accruals and depreciation.
- The final accounts are prepared: a profit and loss account and balance sheet.

A systematic procedure is followed throughout the accounting year in recording transactions. Towards the year end a timetable for aggregating the transactions is followed in order to prepare the final accounting statements. The PLA is compiled at year end (although in larger organizations monthly PLAs are produced and circulated internally) by matching all the costs and expenses by the business in achieving the sales made to customers. The resulting profit (or loss) is transferred to the balance sheet after such distributions as dividends, interest payments and tax. The balance sheet is an expression of the company's financial worth at an instant in time. It is frequently compared with a financial 'snap-shot' of the company at a particular moment. The snap-shot can reveal what was happening at that moment in time. Further, other snap-shots could be taken that might give different impressions. The balance sheet snap-shot is an annual event during the ongoing life of the company.

Further reading

Accounting in a Business Context, by A. Berry and R. Jarvis (Chapman Hall, 1991) takes an interesting approach towards the understanding of accounting using the accounting equation and the matrix approach towards providing final accounts from recording transactions. *Financial Analysis and Control*, by A. Birchall (Butterworth-Heinemann, 1991) and *Accounting for Non-Accounting Students*, by J.R. Dyson (Pitman, 1991) are conventional accounting texts with the emphasis given to double-entry and the basics of financial accounting through a number of examples. *Financial Accounting Methods and Meaning*, by R. Laughlin and R. Gray (Chapman Hall, 1988) is a comprehensive text that adopts a systems approach towards financial accounting while providing more detail concerning SSAPs etc.

Tutorial exercises

6.1 For Brign Engineers prepare a Profit and loss account on the following items:

	£000s
Sales	400
Energy	60
Depreciation	40
Other operating expenses	20
Interest paid	30
Interest received by the firm	40
Administration	20
Wages and Salaries	80

6.2 Slikfit Exhausts Ltd

Liabilities	£000s	Assets			£000s
			At cost	depreciation	
Share capital	100	Fixed	140	40	100
Reserves	20				
Debentures (10 per cent)	40	*Current*			
Trade creditors	12	Stock			35
Dividends payable	10	Debtors			26
		Cash			21
	182				182

The following details relate to Slikfit Exhausts Ltd during the year ended 30 September 19X2:

A Sales totalled £180,000 over the year of which two-thirds of the total were sold on credit and one-third cash.

B Purchases of stocks were made on credit totalling £80,000.

C Depreciation amounted to £20,000.

D Operating expenses incurred for the year were £25,000. £22,000 was paid in cash with £3,000 unpaid at 30 September 19X3.

E The directors declared a final dividend for the year to 30 September 19X3 of £5,000. An interim dividend of £4,000 had been declared during the year and paid. The previous year's dividend was also paid in full during the year.

F Debenture interest of £4,000 was paid in cash.

G The closing stock was valued at £15,000.

H Trade creditors were paid £40,000.

I Debtors were paid £110,000 in cash.

Required

Using the information above, prepare the profit and loss account for the year ended 30 September 19X3 together with a summarized balance sheet at that date.

6.3 Jelly and Partners

The following is a balance sheet at the end of last August 19X3 in a summarized form:

Liabilities	£	Assets	£
Owner's capital	9,000	Machine	4,000
Profit and loss retained	2,000	Stock	2,800
Trade creditors	4,500	Debtors	5,200
		Prepaid rates	1,200
		Cash	2,300
	15,500		15,500

Additional information and details of transactions in September 19X3 are:

• Rates bill for the company is £2,400 p.a. which was paid six months ago.

• Sales amounted to £8,500, of which £1,900 was for cash; the remainder was credit.

• Stocks were purchased on credit for £4,400.

• At the end of the month stocks were checked and had a cost value of £3,200.

• Wages for the month amounted to £2,200, of which £160 had not been paid.

• Trade creditors received £6,100 in cash as part payment of bills outstanding.

• Machine expenses totalled £300 which were settled in cash.

- A bank loan of £7,200 was arranged at 10 per cent per annum and would be paid monthly with the capital being repaid in four years.
- At the beginning of the month £1,200 was paid for business insurance for the year.

Required
Prepare a trading, profit and loss account for September 1993 and a balance sheet as at the end of the month.
The answer is on page 134.

Appendix 6.1: Conventions and concepts in financial accounting

The accounting process of identifying and systematically recording transactions and reporting them in final summarized accounts form is subject to differing legitimate approaches to the way in which measurements are made. The accounting professional bodies attempt to regulate this through the issue of SSAPs (Statements of Standards of Accounting Practice) and, more recently, Financial Reporting Standards (FRSs). However, the diversity of business activity, the nature of differing industries, the varied composition of assets and liabilities, differing interpretations etc. make a rigid system of reporting unwieldy and unclear.

There are, however, a number of concepts and conventions that should underpin any organization's financial accounting system. These concepts and conventions can be considered as principles. The arithmetical manipulations outlined earlier in this chapter are underpinned by the following concepts and conventions. It is the closest accounting can get in providing an overall framework of financial accounting. Essentially, these concepts and conventions are as follows.

Accruals or matching

The determination of profit is arrived at using the concept of accruals in accounting. Profits are determined by matching costs as they are expended and revenues as they are earned (unlike cash flow accounting which recognizes costs and revenues as cash paid or received). Earlier in the chapter, adjustments were made to the accounts to reflect the stock position and depreciation. Both of these are examples of accrual.

Going concern

Any business or organization is assumed, for accounting purposes, to continue to carry out their activities for the foreseeable future. This principle does not attempt to quantify the future. However, it can safely be assumed to extend to future accounting periods. Basically, if the owners of the business were intending to cease trading then interest would focus on the saleable value of the assets, not simply on how well those assets have been used to produce profits.

Consistency

There are a number of differing approaches to depreciation, valuing stocks etc. Consistency means that businesses selecting a particular approach should not alter the method year on year to suit itself. The external parties need to be confident, if comparisons are to be made with previous years, that the methods have not altered, making the comparisons worthless.

Conservatism	Sometimes regarded as the prudence principle. Businesses should not be anticipating profits but ought to recognize potential losses. For example, obsolete stock valued at cost in the balance sheet would suggest that the owners' wealth is greater than it should be. Hence, when there is little doubt that stock has become obsolete or unsaleable, the value of that stock should be written down to its net realizable value (NRV) and the difference charged as an expense to the PLA – thereby reducing the owner's wealth for the period. This is particularly important in bespoke engineering and other businesses where goods are made to order.
Objectivity	This suggests that accounts should be based on measurable facts which can be documented through invoices and receipts and can be independently verified. It avoids non-documented transactions involving 'opinion' and 'judgement'.
Materiality	There are costs associated with the recording and reporting of all transactions in a business. However, the effort spent on trying to ascertain whether a £3 purchase of an office ashtray is treated as capital expenditure, or written off in the PLA, would be pointless given that the latter treatment will make no significant impact on the reported results of the business.
Entity	This concept distinguishes the business activities from the personal affairs of the owners of the business. It is enshrined in the layout of the balance sheet, which separates the ownership of the business from the assets entrusted to the business to manage.
Historical cost	Any assets of the business should be recorded at the price paid for them. This would imply that the value of all assets is equal to the amount paid for them. If there is a fall in value, then the assets should be written down and the difference charged to the profit and loss account. This principle comes into practice when stocks should be recorded at the lower of cost or NRV.
Money measurement	All a company's assets and liabilities are expressed in monetary terms. This enables a whole range of assets such as buildings, machines and stocks, as well as intangibles such as goodwill, to be expressed as a common denominator, i.e. sterling. Recently, firms have tried to place a monetary value on the brands owned by a company.
Duality	This convention follows the principles of double-entry and keeping the accounting equation in balance. It means that for any increase in assets, there must be a commensurate increase in the liabilities. Equally, any reduction in assets will be met by a fall in liabilities. The profit and loss account will reflect this by adjusting the balance sheet owner's equity periodically.
Stability	There is an assumption, for accounting purposes, that the purchasing power of currency remains constant year on year. This is only applied to assist the measurement of assets and liabilities and their verification. Of course, in reality, because of inflation the overall interpretation of accounts

over time must be treated with some caution. The higher the inflation rate the less reliable the results when comparing one set of accounts with previous periods.

Appendix 6.2: The bookkeeping system

It is completely impractical from a logistics and economic perspective, if not impossible, to prepare a balance sheet after every business transaction. The method of recording transactions is known as double-entry. It is a systematic approach to recording all transactions that are quantifiable in financial terms.

In the past, the recording of transactions was undertaken by bookkeepers. In smaller businesses this still tends to be the case, with the likelihood of the bookkeepers being aided by the use of PCs. In large organizations, however, the systematic recording, classifying, aggregating, summarizing and reporting of transactions are undertaken by greater investment in computing facilities. Generally, the larger the organization, the greater the memory capacity and speed required to process large volumes of separate transactions.

For a company the size of ICI, capturing the wide variety of business transactions while operating in different marketing segments (chemicals, animal feeds etc.) with different technologies reflected through many sub-organizations, such as divisions and companies (both home and abroad), will require mainframe computing with bespoke software. In addition, there would be a need for managerial, technical, and clerical support for providing the inputs from documentation, such as invoices, through to producing management reports. In a local electrical retail outlet the recording of most of the transactions will be through the cash register and bank account. The shop owner may be equipped with either their own bookkeeping stationery or PC with commercial software packages such as Sage(TM) or Pegasus(TM).

It is not the intention of this chapter to explain the comprehensive approach to double-entry. The basic principles underlying it will be explained.

Each asset and liability classification will have its own 'T' account:

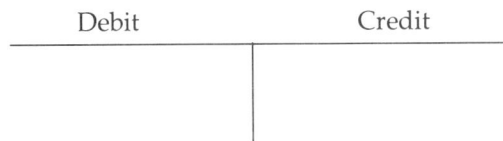

Debit	Credit

The rules are that for any recorded financial transaction two accounts will be affected by the financial sum involved. For example, the provision of capital of £10,000 into the business by the owner in the form of cash will affect two 'T' accounts:

	−	+		+	−
Liabilities:		£10,000	Assets:	£10,000	

131

For every right-hand entry of the transaction in one 'T' account there will be a commensurate left-hand entry in another 'T' account: for every debit entry there will be a credit entry.

The issue however of which account to use is dependent on whether the accounts affected are assets or liabilities. It is important because the left-hand side of each type of account has a different function:

> asset accounts record the increase of that asset on the left-hand side of the account. On the other hand, liability accounts record a decrease in liabilities on the left-hand side.

The converse is true, assets decrease on the right-hand side but liabilities increase.

The above transaction means capital has increased while, simultaneously, so has the asset, cash – *but the double-entry rule has been followed*.

The above diagram makes more sense if this recognizes using '+' or '–' instead of credit or debit:

Liabilities:	–	+	Assets:	+	–
Capital		£10,000	Cash:	£10,000	

But not every business transaction directly affects assets and liabilities. If the next transaction involved converting some of the cash into raw materials, £6,000 then involves converting one asset, cash, into another form of asset, stock, in this case for resale. The double-entry principle will continue: for every left-hand side entry into the 'T' account there will be a commensurate right-hand side entry. Only this time one asset, cash, is *reduced* and another 'T' account is formed to show an *increase* in stock.

Asset:	+	–
	Cash	
	£10,000	£6,000

A casual inspection of this particular account at this point would suggest that the left-hand side (debit) is greater than the right-hand side (credit) by £4,000. This notion takes on greater significance at the end of the accounting period because these accounts would continue recording *ad infinitum* from one period to the next unless they were to be balanced off and an opening balance begins the new period.

Asset:	+	–
	Stock	
	£6,000	

If a balance sheet were to be drawn up at this point in time then the equation will hold:

Liabilities = Assets

Because:

Capital	£10,000	Assets:	
		Stock	£6,000
		Cash	£4,000
	£10,000		£10,000

Earlier we saw that when raw material stocks was converted with the aid of labour two things would happen: first, there would be a reduction of raw material stock and a simultaneous transfer into WIP; second, there would be a reduction of cash equal to the wages and this would be added to the WIP.

In fact, this would not be a satisfactory way of recording and there would be need to be an account opened to record the payment of wages as a transaction. If £2000 'wages' were to be paid then the double-entry rules would follow:

Assets: +   −

	Cash		
	£		£
Capital	10,000		
		Stock	6,000
		Wages	2,000

The new 'T' account will be called wages. It is similar to any asset account except that the wages have been expended in this period and will not benefit any future period. This we will call an Expenditure account:

Expenditure   +   −

	Wages	
Cash	£2,000	

Of course, there will be other expenditures in getting the output of the firm into a finished state and sold. There will be rents and rates to be paid, energy, telephone, salesmen's petrol expenses, heating and lighting, water and so on. All of these items would be given their own separate expenditure accounts.

There are general accounting rules that encapsulate all accounting transactions of any organization:

1 If any asset is increasing or any liability or capital item is decreasing, there is a left-hand side entry or 'debit' in double-entry recording.
2 If any asset is decreasing or any liability or capital item is increasing, there is a right-hand entry or 'credit' in double-entry recording.

A consistent application of the above rules with any transaction guarantees a continuous maintenance of the balance sheet equation without having to produce a balance sheet other than periodically.

Any transaction will need to be analysed by determining its effects on the following:

	Left hand (Debit)	Right hand (Credit)
1 Increase in asset	X	
2 Decrease in asset		X
3 Increase in liability		X
4 Decrease in liability	X	
5 Revenues (always increase)		X
6 Expenses	X	

Note: 1, 4 and 6 always involve double-entry accounts with insertions on the left hand side.

Assets/Expenses

+	−

2 and 3 always involve insertions on the right hand side of the double-entry accounts

Liabilities/Revenues

−	+

Each business transaction will need to be analysed into its component parts. Purchasing stock for cash involves substituting one asset for another which involves 1 and 2 using the table above. Issuing capital for cash involves both 1 and 3. Wages paid involve 2 and 6. Sales on credit involve 1 and 5 whereas using stock for manufacture involves 2 and 6 and so on.

The principles of double entry can also be demonstrated through modelling assets and liabilities on a spreadsheet. This approach mitigates the painstaking 'bookkeeping' approach.

Appendix 6.3: Answers to tutorial exercises

Answer: Slikfit Exhausts Ltd

Slikfit Exhausts Ltd Profit and loss account for the year ending 30 September 19X3

	£000s	£000s
Sales		180
Cost of goods sold:		
Opening stock	35	
+ Purchases of stock	80	
− Closing stock	(15)	
		100
Gross profit		80
Depreciation	20	
Operating expenses	25	
		45

Operating profit	35
Interest on debenture	_4_
Net profit	31
Dividends	_5_
Retained profit to reserves	_26_

Slikfit Exhausts Ltd balance sheet as at year ending 30 September 19X3

Liabilities	000's	Assets	000's		
			At cost	depreciation	
Share capital	100	Fixed	140	60	80
Reserves (20 + 26)	46				
Debentures (10 per cent)	40				
Trade creditors	52	Stock			15
Dividends payable	1	Debtors			36
Accruals	3	Cash			111
	242				242

On the assets side, fixed assets NBV would fall by the £20,000 depreciation for the year.

The new closing stock figure would go to the closing balance sheet.

Opening balance sheet debtors were £26,000. This was increased over the year by credit sales to customers of £120,000. Some of these debtors were reduced by cash being paid of £110,000; therefore closing balance sheet debtors are £36,000.

Opening balance sheet cash was £21,000. This increased by cash in-flows from sales of £60,000, debtors paying cash of £110,000. On the outflow side trade creditors were paid £40,000, interim dividends of £4,000 were paid, last year's dividend on the previous balance sheet of £10,000 was paid, operating expenses of £22,000 were paid, and interest of £4,000 was paid. The remaining cash sum was £111,000 to go in the final balance sheet.

On the liabilities side, opening creditors were £12,000. This increased by purchasing stock on credit over the year by £80,000 although the firm paid £40,000 of this off, leaving £52,000.

The directors declared a final dividend of £5,000. £4,000 of this amount was paid during the year out of cash leaving an obligation to pay the share-holders another £1,000 into the next period (accrual concept). The £10,000 from last year was paid this year out of cash.

The accrual of £3,000 is a current liability because the expenses incurred for profit measurement purposes are £25,000. Only £22,000 was paid out of cash during the year.

Finally, the profit retained of £26,000 is carried to the balance sheet as this is the shareholder's increase in wealth. In fact, the shareholder's profit was £31,000 but £5,000 is to be paid out in dividends, of which £4,000 has already been received.

Answer Jelly and Partners

Profit and loss account for the month ended September 19X3

	£	£
Sales		8,500
Cost of goods sold:		
Opening stock	2,800	
+ Stock purchases	4,200	
− Closing stock	(3,200)	
		4,000
Gross profit		4,500
Rates	200	
Wages	2,200	
Machine expenses	300	
Insurance	100	
Interest	60	
		2,860
Net profit		1,640

(*Note*: Given the simplicity of nature of this business there is no need to present in the company recommended format.)

Balance sheet as at the 30 September 19X3

Liabilities	£	Assets	£
Owners' capital	9,000	Machine	4,000
Profit and loss retained	3,640	Stock	3,200
Bank loan	7,200	Debtors	11,800
Trade creditors	2,800	Cash	1,700
Accruals	160	Prepaid rates	2,100
	22,800		22,800

7 Evaluating final accounts

Final accounts of the various documents, including the balance sheet and PLA, are generated by the organization's transactions and validated by external auditors on behalf of the owners. These final accounts reflect the financial activities of the current accounting period. The accounts are based upon a number of concepts, conventions and specific accounting policies.

The final accounts in themselves are of little use to the reader without undertaking some attempt at analysis. This is undertaken by measuring some key relationships between various items of the balance sheet and PLA. In this chapter, these key relationships are described and calculated, and their significance explained. Figure 7.1 identifies some of the major relationships that would be examined by external parties with vested interests in understanding the financial performance of the organization over the last accounting period. The diagram is not meant to be the definitive version of key relationships; it only serves to demonstrate that there is a link between the various items expressed in the accounts.

Figure 7.1 has been organized into a pyramid format in order to provide some structure to an interdependence of financial relationships expressed in the form of ratios.

At the top of the pyramid is a major profitability ratio used by both internal and external users of financial accounts called the return on capital employed (ROCE). It is sometimes regarded as the primary ratio. It is a percentage measure of the profit generated from the operations of the

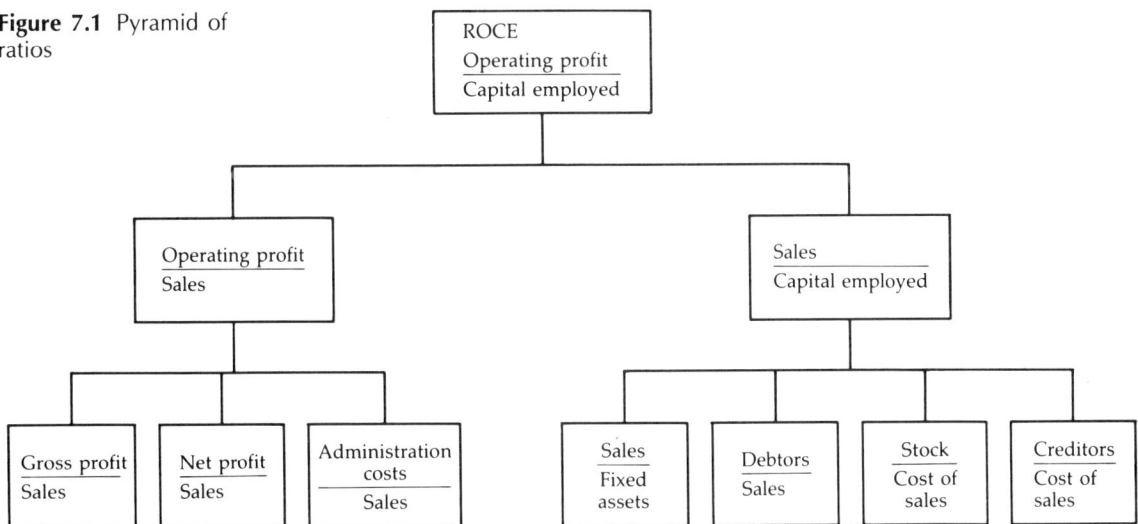

Figure 7.1 Pyramid of ratios

137

organization over a specific time period (usually one year if calculated from published accounts) divided by the capital employed in generating that same profit over the same year.

$$ROCE = \frac{\text{Profit from operations before interest}}{\text{Capital employed}} \times 100$$

There are a number of refinements that can be made to the definition of ROCE. The underlying logic however is determined by the relationship of the numerator measured against the denominator, and a consistent approach to measuring against other competitors and/or over a number of time horizons for the same organization.

Capital employed

Capital employed is a measure of the financial resources used to generate the profits or losses of the business. It has two main constituent parts, shareholders' capital and loan capital. Shareholders' capital is contributed by the original owners when the shares were first issued. At the time of issue, each share would have had a nominal face value, but is likely to have been purchased at a considerably higher price than this nominal value. The difference between actual issue price and the nominal value is referred to as share premium, and is recorded separately in the accounts. Any reserves, from past profits not distributed to shareholders, or from revaluing assets which have appreciated, also belong to the shareholders and are considered as shareholders' capital.

Loan capital is provided by individuals and organizations external to the firm (although owners can lend money to their own company in a private capacity). Such loans would include mortgages on property, bank loans and debentures; the latter being a particularly safe kind of loan which normally pays a fixed rate of interest, is secured on the company's assets and can normally be sold on by the lender to another party.

For a working definition of capital employed, both share capital and loans (or creditors) beyond one year are included. The reward for each constituent element of capital is dividends and retained profits for the shareholders and, for lenders, interest payments. Profit from operations is the collective sum to reward both categories of capital provider, although lenders expect to receive interest regardless of whether the firm has made any profit or not.

Shareholders capital =	Issued shares (nominal) + Share premium + Reserves
Loan capital =	Debentures + Loans beyond one year + Mortgages

Figure 7.2 Composition of capital employed

From a balance sheet perspective, the terminology is important – but each heading will be subject to the original accounting equation of Assets = Liabilities:

Capital employed = Net assets
Net Assets   = Fixed assets + (Current Assets – Current Liabilities)

Occasionally, the ratio is expressed as Return on Net Assets (RONA), but since capital employed and net assets will always turn out to be the same thing,

ROCE = RONA

If current liabilities were added to capital employed then the original accounting equation of Assets = Liabilities is maintained.

Using ROCE

The ROCE therefore, can be considered as a financial productivity ratio because it measures the reward to long-term capital providers, i.e. profit before interest, expressed as a percentage of the total capital at the last balance sheet date.

From a manufacturing management perspective, productivity can be regarded as the outputs of a system divided by the inputs. Expressed in this manner productivity is said to improve if the resultant number increases over time:

$$\text{Productivity} = \frac{\text{Outputs of system}}{\text{Inputs of system}}$$

But as with any technologist responsible for improving productivity there are a number of options available. From a financial perspective there is a range of methods that senior management can use to improve the ROCE productivity measure:

1 Increase the profit from operations by increasing sales revenue and reducing costs, while maintaining the same capital employed;
2 Maintain current profit level, while reducing the capital employed through, for example, reducing loans or buying back shares;
3 Combine 1 and 2 above;
4 Increase the profit from operations at a greater rate than an increase in capital employed;

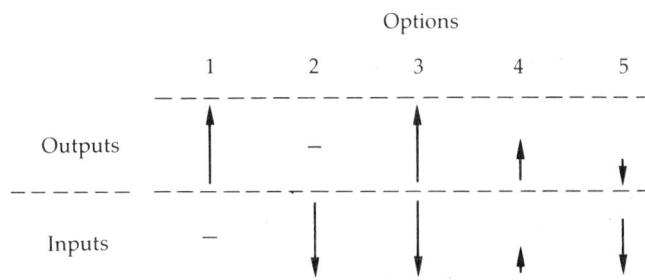

Figure 7.3 Alternatives for increasing financial productivity

5 Reduce the profit from operations but at a slower rate than reducing the capital employed through, for example, selling off surplus capacity.

The ROCE, then, allows analysts and senior management to look at the business as a whole. The ROCE figure for a particular firm can be compared with previous performance, competitors and budget targets (see Chapter 9). Although ROCE is extremely important, a deeper analysis requires that other relationships be considered.

Secondary ratios

The second tier of the pyramid in Figure 7.1 is:

$$\frac{\text{Operating profit}}{\text{Sales}} \times 100 \text{ and,} \quad \frac{\text{Sales}}{\text{Capital employed}}$$

Both ratios are a further breakdown on the primary ROCE ratio and are referred to as secondary ratios. If the two ratios are multiplied together then the result will be the ROCE expressed as a percentage. This relationship between the primary and secondary ratios suggests that, if the ROCE is to improve, either the return on sales or the turnover of capital needs to increase. The former is known as the Return on Sales (ROS) ratio and the latter the Asset Utilization (or Capital Turnover) ratio. Sometimes the Capital Turnover is expressed as Sales/Net Assets and called Asset Utilization. As the accounting equation holds, the resultant figure is identical to sales/capital employed.

Operating profit is used as the numerator of the ROS ratio, not simply net profit. The reason is that net profit can be calculated after interest has been deducted. Interest needs to be added back to give the higher operating profit figure as the numerator, since the capital contributed by long-term lenders is included in the denominator. If interest is removed from the profit figure, and loan capital deducted from the denominator, the resultant ratio shows the return on shareholders' funds only; itself a useful ratio for shareholders.

The remainder of the pyramid in Figure 7.1 is simply one of expressing further financial relationships from the financial statements.

Profit to sales ratios

For example, the sales revenue figure in the PLA gives an indication of the purposeful activity of the firm's ability to sell its products and/or services. It is useful to compare this figure with the gross profit generated by sales, by way of a ratio, and contrast the ratio with that achieved in previous years and that achieved by competitors.

$$\frac{\text{Gross profit}}{\text{Sales revenue}} \times 100$$

Of course, the revenue figure is a combination of differing products, volumes and a range of prices. Equally, the same could be said of the cost of sales figure which has been aggregated by the accounting process.

Nevertheless, the overall percentage arrived at indicates the average returns achieved on sales of getting the product/service into its final state ready for sale. The ratio is even more meaningful when used in conjunction with the next ratio: that of net profit to sales.

$$\frac{\text{Net profit}}{\text{Sales}} \times 100$$

This would indicate the net profit achieved on sales after taking into account other functional costs such as selling, marketing, administration and finance costs (such as interest). These are often grouped together under the heading of overheads.

The ratios examined so far will help managers to predict the effects of business activity on the balance sheet. For example, if a manufacturing company wishes to improve its profits, it could achieve this by heavy capital expenditure. However, such expenditure will reduce the ROCE figure in the short-term, which might be taken to herald a worsening of the company's fortunes, when the reverse may be the case. Alternatively, a policy to increase sales might suggest greater credit limits could be offered to customers. This may improve the gross profit to sales ratio considerably. However, the net profit to sales figure might decrease in the following accounting period, if some of the customers never settle their bills. However, if either (or both) strategies were tried successfully, the ROCE figure would ultimately improve. Even so, the company could run into a cash shortage, referred to as a liquidity problem.

Liquidity

Liquidity is the ability of a business to meet its external cash obligations as they fall due. When creditors expect to be paid the business needs to ensure there is sufficient cash to meet this obligation. If resources are all tied up with stocks that have not been sold, or debtors who will not pay up, the firm may come under increasing financial pressure. The techniques for reducing the risk of this happening are known as cash flow forecasts, and are examined in Chapter 9.

It is vitally important – for small businesses particularly – to realize that making a profit, even a good profit, will not guarantee the success or survival of a business in the short term. Failure to generate sufficient cash to meet obligations has been a prime cause of business failure in the UK. Firms in the middle of expansion programmes, when working on large scale engineering and construction contracts, are particularly at risk. Liquidity concentrates on the short-term ability to convert stocks into cash.

The elements in Figure 7.4 are expressed periodically in the balance sheet as current assets and current liabilities. Working capital is the difference between current assets and current liabilities. The flow is important to maintain because any breakdown will affect cash directly or indirectly. It is the responsibility of accountants to ensure that frequent reports are provided internally and management controls exercised on the cycle and its constituent elements. For periodical analysis there are a number of key ratios measuring the liquidity position:

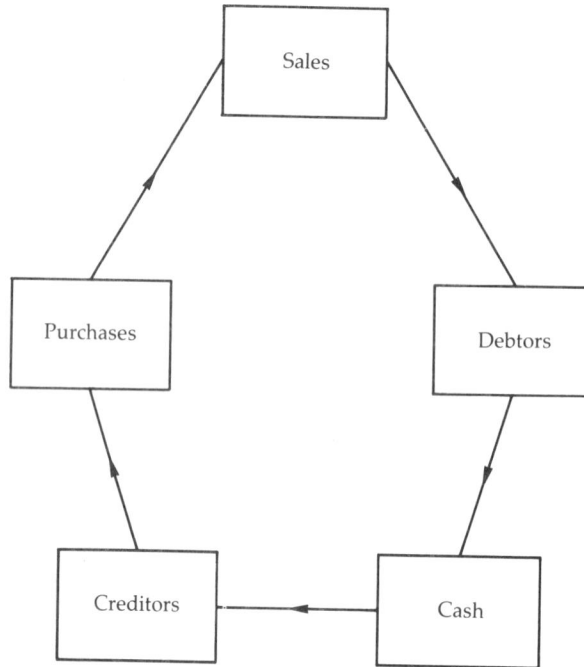

Figure 7.4 Working capital cycle

The current (or working capital) ratio:

Current assets
—————————
Current liabilities

This ratio indicates the firm's ability to meet short-term debts as they fall due. Broadly, if the numerator is greater than the denominator, it implies there are more current assets than current liabilities and may indicate an ability to pay short-term debts. The opposite could indicate a liquidity problem. The norm cannot be stated without benchmarking against other firms, industry norms, and previous years.

A further refinement of the above ratio would be to reduce current assets by the stock value. This is known as the 'acid-test' or 'quick' ratio and might prove to be more significant if the business invests in substantial stocks that may not be readily saleable if debts fall due:

Current assets – stock
————————————
Current liabilities

Construction contractors, by the nature of their business, carry large stocks and WIP. For example, if housing construction companies cannot sell newly built properties, then their stocks will not be regarded as very liquid.

A number of other ratios can be examined, to provide further illumination on liquidity, and where problems may be occurring. Using Figure 7.4 above, each component can provide its own ratio.

If the debtors in the balance sheet are assumed to be a reasonable reflection of sales throughout the year, then by expressing it as a proportion of sales or turnover figure given in the PLA, an indication will be given as to how significant they are in terms of total sales for the year. Further, if the ratio is multiplied by twelve, the answer will indicate in months just how quickly debtors turn over in a year.

$$\frac{\text{Debtors}}{\text{Sales}} \times 12 = \text{Debtor months}$$

(If the above ratio is multiplied by 365 instead then the expression will be in debtor days.)

Debtors are expressed as a proportion of sales because both items are determined by the outputs of the firm through prices charged to customers. This is in marked contrast to the creditors' turnover ratio below. The relationship is expressed as creditors as a proportion of cost of sales (or better still, 'purchases' if the published information is available) because they are both inputs to the firm and should be regarded at cost.

Creditors' ratios are expressed as a proportion of the cost of sales and can be expressed as a simple index or in days/months.

$$\frac{\text{Creditors}}{\text{Cost of sales}} \times 12 = \text{Creditor months}$$

Finally stock turnover can be expressed as a proportion of the cost of sales and indicate the number of times in a year, days and months, stock is converted into sales. Again, because it is expressed 'at cost' in the balance sheet it is important to be consistent with the total cost of sales over the year:

$$\frac{\text{Stock}}{\text{Cost of sales}} \times 12 = \text{Stock turnover}$$

If there is a consistent expression of creditors, stocks and debtors it is possible to gain some insight as to how quickly the assets and liabilities can be converted into cash. This is called the cash conversion cycle:

	Stock months
add	Debtor months
less	Creditor months
=	Cash conversion cycle in months

Internal financial management requires the cycle to be monitored constantly. A deterioration in the cycle will indicate an increasing risk of liquidity problems.

Solvency and investor's ratios

There are other important relationships between items in the final accounts. One particular ratio that will be briefly mentioned here is the Gearing ratio. This is found by examining the composition of capital, broadly expressed in terms of debt and equity, in the balance sheet.

143

$$\frac{\text{Debt}}{\text{Equity}} \times 100 \quad \text{or} \quad \frac{\text{Debt}}{\text{Debt} + \text{equity}} \times 100$$

The term debt/equity ratio is often used instead of gearing. The importance of this ratio will be further developed in Chapter 11, but briefly, the greater the resulting ratio, the more pressure there is on the company to make profits in order to meet interest payments. Failure to meet these interest payments could result in the liquidation of the company.

A worked example of ratio analysis is given in Case Study 7.1, TriEngines, below. Like a large number of engineering organizations TriEngines is a Plc, and, as a consequence, follows reporting formats laid down by the accounting professions. All accounting annual reports will show the profit and loss accounts of the current year's operations and balance sheet. These figures are also compared with the previous year's PLA and balance sheet. In addition there are other reports provided such as the Chairman's statement, funds flow statements and notes to the accounts. For example the notes to the accounts may indicate the composition of stock or purchase and disposal of assets.

For the purpose of the book it should be pointed out that the only acceptable presentation of the balance sheet is the

Assets = Liabilities

style referred to earlier. The balance sheet of TriEngines is abridged and gives less detail than would be found in a proper set of annual accounts.

Case study 7.1: Tri Engines

Abridged balance sheet for TriEngines Plc as at 31 December 19X1

	(£000's)	(£000's)	(£000's)
Fixed assets			
Land and building		800	
Plant and machinery		600	
Fixtures and fittings		200	
			1600
Current assets			
Stock	600		
Debtors	300		
Cash	250		
		1150	
Current liabilities			
Trade creditors	470		
		470	
Net current assets			680
Net assets			2280

Capital and reserves

Issued ordinary share capital (£1 nominal)	700	
Reserves (includes 401 retained, see below)	900	
Shareholders' interest		1600

Creditors beyond one year

Debentures		680
Capital employed		2280

Profit and loss account for TriEngines Plc for year ended 31 December 19X1

		(£000's)
Sales turnover		4200
Less: Cost of sales		2500
Gross profit		1700

Less expenses:

Distribution	110	
Administration	470	
Debenture interest	84	
		664
Net profit before tax		1036
Less: Tax		460
Profit after tax		576
Less: Dividends		175
Transfer to retained earnings		401

Abridged balance sheet of TriEngines Plc as at 31 December 19X2

		(£000's)
Fixed assets		
Land and building	970	
Plant and machinery	660	
Fixtures and fittings	300	
		1930
Current assets		
Stock	700	
Debtors	450	
Cash	380	
	1530	
Current liabilities		
Trade creditors	540	
	540	
Net current assets		990
Net assets		2920

Capital and reserves			
Issued ordinary share capital (£1 par)	750		
Reserves	1270		
Shareholders' interest		2020	
Creditors beyond one year			
Debentures		900	
Capital employed			2920

Profit and loss account for TriEngines Plc year ended 19X2

	(£000')	(£000's)
Sales turnover	5460	
Less: Cost of sales	3500	
Gross profit		1960
Less expenses:		
Distribution	143	
Administration	505	
Debenture interest	112	
		760
Net profit before tax		1200
Less: Tax		480
profit after tax		720
Less: Dividends		350
Transfer to retained earnings		370

Using the above two sets of accounts then there are some key relationships that needed to be examined. The financial productivity measure of ROCE would be a convenient starting point.

Profitability

$$\text{ROCE} = \frac{\text{Profit from operations before interest}}{\text{Capital employed}} \times 100$$

19X1	19X2
$\dfrac{576 + 84}{2280} \times 100 = 28.9 \text{ per cent}$	$\dfrac{720 + 112}{2920} \times 100 = 28.5 \text{ per cent}$

It can be seen from this key overall business performance ratio that performance has slipped from 33.3 per cent in 19X1 to 28.5 per cent in 19X2. Profits have increased but then so has the capital employed, necessary to generate the additional profits.

In order to investigate this matter further the asset utilization (or capital turnover ratio) will be used to determine whether sales improved as a consequence of increasing assets.

(£000's)

	19X1		19X2	
Sales	4200		5460	
Capital employed	2280	= 1.84	2920	= 1.87

This ratio is not usually expressed in terms of percentage. This would suggest there has been hardly any change in the additional sales generated by a proportional increase in the capital employed. Consider the return on sales ratio:

$$\frac{\text{Operating profit}}{\text{Sales}} \times 100: \quad \frac{760}{4200} \times 100 = 18.1\% \quad \frac{832}{5460} \times 100 = 15.2\%$$

This would appear to provide a greater indication as to why the overall financial productivity figure fell. Operating profits increased but not enough to keep up with sales. The operating profit in this case is the profit figure after tax but adding back the interest to enable a consistency with the ROCE at this stage. The next stage would be to identify from the final accounts where the problems have arisen. If the trading profit proportion is considered first:

(£000's)

	19X1		19X2	
Gross profit	1700		1960	
Sales revenue	4200	×100 = 40.4%	5460	×100 = 35.9%

$$\frac{\text{Gross profit}}{\text{Sales revenue}} \times 100: \quad \frac{1700}{4200} \times 100 = 40.4\% \quad \frac{1960}{5460} \times 100 = 35.9\%$$

There has been a fall in the gross profit margin.

As gross profit is the difference between sales and cost of sales a number of engineering and management activities can be evaluated externally. Sales have increased but the cost of sales has increased at a greater rate. Pricing to customers and the terms of trade (contract, discounts) may have to be reviewed. In addition, the purchasing and stock-holding policies should also be examined.

Net profit ratio

(£000's)

$$\frac{\text{Net profit after tax}}{\text{Sales revenue}} \times 100: \quad \frac{576}{4200} \times 100 = 13.8\% \quad \frac{720}{5460} \times 100 = 13.2\%$$

Net profit as a proportion of sales has also fallen marginally over the year. This is only a partial reflection of the earlier gross profit margin fall and would suggest that performance in the other aspects of the business may have improved (such as administration and distribution) as a proportion of sales.

$$\frac{\text{Admin. costs}}{\text{Sales}} \times 100: \quad \frac{19X1}{4200} \times 100 = 11.1\% \quad \frac{19X2}{5460} \times 100 = 9.2\%$$

There are other relationships that could be determined to assist the analysis depending on the details provided in the profit and loss account.

Liquidity

The next stage of the analysis would be to consider the position of the firm's liquidity. The current assets and liabilities change at a greater rate over a measurement period than do fixed assets and capital.

$$\frac{\text{Current assets}}{\text{Current liabilities}}: \quad \frac{19X1}{470} = 2.4 \quad \frac{19X2}{540} = 2.8$$

Generally, it is regarded that should current assets be greater than liabilities then the firm could meet its short-term obligations as they fall due. In this case there has been an improvement in liquidity.

There is a further refinement of this ratio that ought to be considered should stock be a significant part of the business operation. In engineering construction for example there would probably be a large proportion of WIP. The 'quick' or 'acid-test' ratio removes stock from the calculation if it was considered that stock is not sufficiently liquid. It is worth considering that no matter how much has been spent on a project, if cash is not recovered sufficiently quickly, creditors may pressure payment from the firm's current assets and the WIP may not prove to be anywhere near saleable or even likely to recover the costs expended to date.

$$\frac{\text{Current assets} - \text{stock}}{\text{Current liabilities}}: \quad \frac{550}{470} = 1.17 \quad \frac{830}{540} = 1.53$$

The ratio after removing stock suggests that there are sufficient liquid assets to meet liabilities. Again, this would include debtors and if the company were exposed to major customers that provided a high proportion of sales then this factor too would have to be taken into account.

A further breakdown can be made concerning liquidity by examining the elements of the working capital cycle (current assets less current liabilities). Debt customers appear to be paying promptly in this business but the debtors ratio is beginning to fall, i.e. debtors are taking longer to settle.

	19X1	19X2

$$\frac{\text{Debtors}}{\text{Sales}} \times 365 = \text{Debtor days:} \quad \frac{300}{4200} \times 365 = 26 \quad \frac{450}{5460} \times 365 = 30$$

Both numerator and denominator are expressed in sales terms for consistency.

On the other hand, it appears that the firm is paying its suppliers much earlier than it did in the previous year.

	19X1	19X2

$$\frac{\text{Creditors}}{\text{Cost of sales}} \times 365 = \text{Creditor days:} \quad \frac{470}{2500} \times 365 = 68 \quad \frac{540}{3500} \times 365 = 56$$

Both numerator and denominator are expressed in cost terms for consistency.

The turnover of stock is increasing. More simply, the throughput of raw materials to WIP to customers is increasing by it being held in the firm for fourteen days less than the previous year.

	19X1	19X2

$$\frac{\text{Stock}}{\text{Cost of sales}} \times 365 = \frac{\text{Stock turnover}}{\text{(in days)}} \quad \frac{600}{2500} \times 365 = 87 \quad \frac{700}{3500} \times 365 = 73$$

Again, both stocks and cost of sales are expressed in cost terms and should be consistent in the ratio. This ratio could be refined by expressing stocks as a proportion of purchases if the information was provided externally.

The cash conversion cycle can be calculated. If debtors are taking four days longer to pay than last year and creditors are being paid twelve days faster, then the ability to convert into cash is reduced by sixteen days. But if stocks are turned over faster than last year by fourteen days then the cash conversion has fallen by a net two days on last year.

	19X1	19X2
Stock days	87	73
add Debtor days	16	20
less Creditor days	(68)	(56)
= Cash conversion cycle in days	35	37

Solvency

There is a link with short-term liquidity and the long-term notion of solvency. Solvency, or gearing, ratios attempt to measure the exposure the firm may have to financial risk caused by the composition of the capital structure in the balance sheet. If the firm carries a lot of debt relative to equity (i.e. share capital) in its balance sheet it would mean that a large proportion of profits must be paid to the debt providers in the form of interest before dividends can be made to shareholders. Should profits be growing at a greater rate than the debt, this will benefit the shareholders:

$$\frac{\text{Debt}}{\text{Equity}} \times 100: \quad \frac{19X1}{\underset{1600}{680}} \times 100 = 42.5\% \quad \frac{19X2}{\underset{2020}{900}} \times 44.6\%$$

In this case the balance sheet value of the debt has increased over the year by more than equity, but not significantly so. There are other forms of gearing ratio but all attempt to measure the vulnerability of the firm's capital structure (composition of debt and equity) relative to changes in business activity as measured through sales and profits.

Overall the company appears to be in no immediate danger when examining the above ratios. However, the company has deteriorated in profitability terms from last year for reasons that would need to be investigated and controlled by management. Its liquidity appears to be satisfactory, but there would need to be consideration of why creditors were being paid so much earlier.

Limitations of financial accounts

The external output of financial statements will require a close examination and evaluation of the organization if some understanding of its overall performance is to be gained. The engineer may be a consumer of these accounts from a prospective supplier or customer perspective. The engineering manager may be considering a potential acquisition. Any evaluation can only be tempered by an appreciation that the information contained therein must be followed up by further investigation

In practice the above example of TriEngines Plc could be considered restrictive because there were only two sets of abridged accounts to compare. Published reports of companies need not provide a great deal more information than current year and previous year financial statements together with cash and funds flow statements and notes to the accounts.

Therefore the above analysis should be regarded as being rather limited and in need of further investigation. This sort of analysis should not be regarded as anything other than an initial 'financial screening' that cannot conclude or recommend without consideration of the following points.

1 Further background information

The nature of the industry and its major competitors should also be appraised.

Two years' comparison may not be sufficient and a five-year comparison may be more information (although there may need to be some inflation adjustments).

2 Future

The economic and marketing forecasts for the industry.

Budgets, plans and projections of the firm's future earnings potential.

Information concerning major initiatives, contracts and investments that will have a direct bearing on the future revenue stream of the business.

3 Valuations

There are a number of accounting policies that may affect the published results of the company at any one point in time. This information would have to be established from the notes that accompany a set of final accounts concerning such as stocks and depreciation.

Buildings and land would be shown in the balance sheet at cost. However, in the 1980s there were substantial increases in property prices. This made companies very uncomfortable because the balance sheet might be substantially understating its worth to the shareholders and other interested parties. The reverse is the care when property values slump.

4 Management

Increasingly there is a demand to provide greater details about the directors and senior managers of a business in terms of pay and shareholding.

Information concerning budgeting and control systems.

5 Creative accounting

Creative accounting has become both a fashionable (particularly in the 1980s) and rather misunderstood term. Quite simply, it involves accountants presenting objective financial results in such a way that conventions, laws and professional standards are not breached, and yet a positive impression is given. The scope of this book does not allow a lengthy discussion on the methods of being 'creative' in accounting. But, the reader should be aware that different acceptable methods of, for example, depreciation and stock valuation stated earlier in this chapter can give rise to a potential range of differing profits and balance sheets within the same time period. Moreover, a favourable impression can also be obtained by including, for example, such special one-off receipts from, say, the company pension fund, to the profit and loss account, thereby inflating the profit figure. This might imply, to a casual reader of the accounts, that the trading position of the firm was stronger than actual performance would justify. A more experienced analyst would carefully study the many footnotes appended to the accounts, and, by making adjustments for any creative items, be able to improve the focus of the analysis and make informed judgements.

Case study 7.2: Spring Ram

By Jane Fuller

Spring Ram, the kitchens and bathrooms group, finally brought out its annual report and accounts yesterday, with several more items under the 'notes to the accounts' heading.

The document was read with particular interest by analysts, because the group blamed its sharp fall in 1992 profits – the first setback since its flotation in 1983 – on a change in accounting policies and pressure for a conservative approach.

The report includes a proposal that Arthur Andersen be reappointed as auditors on 21 May.

The board had engaged in much haggling with the accountancy firm prior to bringing out a pre-tax profit figure of £26.2m, a third less than expected.

The extra work earned Andersen higher auditing fees of £170,00 (£140,000) and an additional £391,000 (£136,000) on non-audit fees.

A few of the new notes, however, were criticized by analysts for not showing 100 per cent conservatism: for instance, the carrying forward of some marketing and business development costs and the capitalizing of interest on some investment in fixed assets.

There was also quibbling over the group's claim to having year-end net cash of £10m, because the figures excluded £4.74m owed in non-bank loans and £6.52m in bills of exchange.

Mr Stuart Greenwood, the finance director, who has resigned over accounting controversies at the group but is carrying on until a replacement is found, said these items were partly to do with tax and partly trade credit.

On the accounting policy questions, he commented: 'Some would say not everything is conservative, but I think the accounts are substantially prudent.'

Financial Times 23 April 1993

Creative accounting is rarely illegal, and, to a large extent, is limited by the rules and conventions of accounting practice listed in Appendix 6.1. However, the ability of accountants to manipulate the final accounts in this way makes professional analysis significantly more sophisticated than implied in this chapter.

Chapter summary

By studying the final accounts of a company, much can be learned about its profitability, liquidity and solvency, using ratio analysis. These ratios are most useful when contrasted with those for the same company in earlier accounting periods and its competitors in the same industry. It is also possible to contrast the ratios with industry averages by using such sources as Dunn and Bradstreet. However, caution must always be taken with such analysis, creative accounting can easily mislead an unwary analyst. Case

Study 7.3 in Appendix 7.1 is a ratio analysis and interpretation of Pilkington Plc. In this case, the analyst has adjusted the accounts as part of the analysis, and therefore it is not possible to recreate all the ratios given by referring to the published accounts.

Further reading The texts sited in Chapter 6 offer guides to ratio analysis.

Tutorial exercises

7.1 (a) Appendix 7.1 shows an extract of the 1991 published accounts for Rolls-Royce Plc. The full accounts provide additional information to the profit and loss accounts and balance sheet.

Using the ratios identified above, examine Rolls-Royce's overall performance when compared with the previous year.

(b) Appendix 7.2 also contains a case study on Rolls-Royce, taken from the *Financial Times*. Do interpretations of your ratio analysis match those of the *FT* article?

(c) What strategic alternatives are open to Rolls-Royce?

7.2 The balance sheet for Sticky Sweets, (see Chapter 4, is given below) together with other information. Using ratio analysis, show whether or not the financial state of the company reflects its operational difficulties.

In year 19X1, Sticky Sweets made a profit of £100,000 on a turnover of £2,000,000. Encouraged, Mr Sticky began a major expansion programme, intending to reach a turnover of £3,000,000 by 19X3. Most of the required capital expenditure was made in 19X2, and Research and Development into a new kind of confectionery product was begun immediately.

The turnover for 19X2 reached its target of £2,500,000, but profits were rather disappointing at £130,000.

Sticky Sweets Ltd.	*As at 1 April 19X1 £000*		*As at 1 April 19X2 £000*	
Fixed assets (nbv)		500		630
Current assets				
Raw materials	300		370	
Finished goods	600		350	
WIP	55		103	
Debtors	295		440	
Cash	5		2	
	1,255		1,265	
Current liabilities				
Creditors	450		470	
Bank overdraft	170		190	
Proposed dividend	2		4	
	622		664	
Net current assets		633		601
Net assets		1,133		1,231

Capital

10,000 Ordinary shares at £2	20		20	
Debentures	380		380	
Reserves	633		701	
Transfers from PLA	100		130	
Shareholders' funds		1,133		1,231

7.3 Sticky Sweets has two main competitors. Their balance sheets are given below. Use ratio analysis to assess the strengths and weaknesses of each. Do any solutions to Sticky's problems suggest themselves?

	Sharp Sweets *£000*	*Spicy Sweets* *£000*

Profit and loss account for year ended 1 April 19X2

Sales	2,400	2,900
Less cost of sales	1,800	2,320
Gross profit	600	580
less overheads	400	420
Net profit before tax	200	160
Tax	80	60
Dividend	30	40
Retained profit for year	90	60

Balance sheet as at 1 April 19X2

	£000		*£000*	
Fixed assets at NBV	800			1,035
Current assets				
Stock	550		320	
Debtors	600		500	
Cash	50		5	
	1200		825	
Less current liabilities				
Tax	80		60	
Creditors	430		640	
Overdraft	–		550	
	510		1,250	
Net current assets		690		(425)
Long-term liabilities				
8% Debentures		–	200	
Net assets		1,490		410
Share capital		1,000		300
Reserves		490		110
Net capital Employed		1,490		410

Appendix 7.1

Case study 7.3: Pilkington Glass

Ratio analysis and interpretation of Pilkington Glass

This appendix contains the results of a rigorous analysis of the published accounts of Pilkington Plc. Some of the ratios have been adjusted or 'improvised' to take into account some of the particular activities of the company; redundancy for example.

Pilkington plc: Ratio calculations

Key ratio		1992	1991	1990	1989	1988
Gross profit %		23.79	24.19	27.36	24.72	24.69
Pretax	BRC	4.37	5.72	11.11	12.93	13.42
profit &	ARC	2.95	5.72	10.78	12.64	12.96
After	BMI	1.34	3.29	7.21	8.81	8.67
tax profit %	AMI	0.36	2.43	6.37	7.81	7.74
	AMI & EI	1.04	0.60	6.13	6.86	7.49
After tax profit as a % of	PMI	2.03	4.90	12.13	15.21	15.51
fixed	AMI	0.55	3.61	10.71	13.49	13.85
assets	AMI & EI	1.57	0.89	10.31	11.85	13.41
After tax profit as a % of	PMI	3.05	7.46	16.16	22.52	20.32
shareholders'	AMI	0.83	5.50	14.27	19.96	18.15
funds	AMI & EI	2.37	1.35	13.74	17.54	17.57
Earnings per share		1.2p	8.6p	25.0p	27.3p	26.2p
Dividends per share		6.0p	10.5p	10.5p	9.5p	8.4p
Current ratio		1.89:1	1.65:1	1.77:1	1.69:1	1.87:1
Quick ratio (acid test ratio)		1.17:1	0.95:1	1.03:1	0.96:1	1.13:1
Gearing		80.07%	79.51%	61.15%	76.77%	54.05%
Debt ratio		0.60:1	0.61:1	0.57:1	0.61:1	0.57:1
Return on capital employed	PMI	1.56	3.88	9.17	11.63	11.33
	AMI	0.42	2.86	8.10	10.31	10.12
	AMI & EI	1.21	0.70	7.80	9.06	9.80
RPI inflation to end of March		4.03%	8.24%	8.10%	7.88%	3.48%

BRC	Before redundancy costs
ARC	After redundancy costs
PMI	Pre-minority interests
AMI	After-minority interests
AMI & EI	After-minority interests and extraordinary items

Pilkington: Key financial figures for five years ended 31 March 1992

	1992 £million	1991 £million	1990 £million	1989 £million	1988 £million
Turnover	2,611.0	2,649.5	2,915.0	2,573.0	2,333.0
Gross profit	621.2	641.0	797.5	636.0	576.0
Profit before tax					
BRC	114.0	151.6	324.1	332.6	313.2
ARC	77.0	151.6	314.3	325.2	302.3
Profit after tax					
PMI	35.0	87.2	210.2	226.6	202.2
AMI	9.5	64.3	185.6	200.9	180.5
AMI & EI	27.2	15.8	178.7	176.5	174.8
Total fixed assets	1,727.0	1,779.7	1,733.3	1,489.8	1,303.7
Total shareholder funds	1,147.0	1,169.6	1,301.0	1,006.3	994.7
(Includes MI)					
Current assets	1,107.1	1,182.4	1,286.5	1,121.4	1,032.4
Debtors and cash	683.9	680.3	751.2	638.0	625.0
Current liabilities	585.1	714.7	728.3	663.4	552.1
Total borrowings	918.4	929.9	795.6	772.5	537.7
Total liabilities and provisions	1,687.1	1,792.5	1,718.8	1,604.4	1,341.1
Total assets	2,834.1	2,962.1	3,019.8	2,611.2	2,336.1
Total assets less current liabilities	2,249.0	2,247.4	2,291.5	1,947.8	1,784.0

Pilkington Plc:

1 Profitability

Ratios over five years: Gross profit rising to 1990, fall in 1991 and 1992
  PBT per cent rising to 1988, falling 1989–1992.
  ATP per cent rising to 1989, down in 1990–1992.
  ATP per cent of FAs rising to 18989, down in 1990–1992.
Results indicate the effect of the recession upon an industry that relies heavily upon the construction and automobile industries for its sales, both of which were in severe recession.

Possible reasons for fall in GP per cent:

• Fall in turnover 1992 on 1991 across most areas of the world has now gone down to only 1.45 per cent coupled with a declining fall in profitability.
• To maintain market share, prices may have been reduced; fall in selling prices may now be over, suggesting improved market trading conditions.
• Rises in raw material prices which cannot be passed on to the customer due to the adverse market conditions.

- Fixed price contracts left over from earlier good years which cannot now be altered to give a more realistic selling price.
- Drop in efficiency when running plants at lower than optimum capacity leading to higher unit costs for each product.

Fall in turnover 1.45 per cent, fall of 25 per cent on PBTBRC, 49 per cent after redundancies and decline of 59 per cent in ATP.
Falls due to:

- Rise in distribution costs of 7.11 per cent despite the fall in turnover.
- Higher interest charges:
  Borrowings down 1.25 per cent, interest up 6.60 per cent but interest capitalized down from £8.4M to £0.2M.
- Interest receivable down 29 per cent although cash on balance sheet has risen; this fall is probably due to falling interest rates world wide.
- Fall of 1.56 per cent in wages and salaries and fall, too, in directors' emoluments by 4.45 per cent together with a decrease in pension contributions. This is due to there being one less director this year, rather than a new-found reluctance to take large pay increases. The real rise in directors' remuneration was 5.1 per cent.
- Overseas taxation has taken a heavier toll on profits than UK corporation tax, being 39 per cent this year.

Fall in fixed admin. costs has contributed to reduce the fall in PBT and ATP per cent:

- Continuing rationalization of the group has now resulted in £37M re-construction exceptional item together with a large one off bad debt of Flachglas.
- Slowdown in fall in profitability bears out the claims of rigorous cost reduction.

2 Performance
EPS rising up to 1989, small fall in 1990 and large decline in 1991 and 1992.

Dividends maintained in 1991 in order to keep confidence in the share price and keep this up, but revised dividend policy this year has led to a cut in the dividend in order to retain funds for future growth and keep dividends at a sustainable level. Part of the fall in DPS this year arose from the issue of a number of new shares.

ATP per cent and ROCE per cent rising up to 1989, small fall in 1990 and again large decline in both 1991 and 1992.

Performance due to poor market conditions and the failure of an industrial company to maintain turnover, productivity and profitability in tough times.

3 Liquidity
Both current and quick ratios have improved in the current year, and as current ratio is well above 1 and quick ratio is now above the 1:1

level, there is currently no cause for concern. However, further declines in liquidity and profitability combined might cause creditors to worry.

Debt ratio: around highest level at the end of the year and staying at a level of more than 0.50:1, therefore business is traditionally dependent on its creditors for some of its financing, but the ratio is currently not so high as to cause concern to creditors that they would not get paid in full.

Gearing: constantly rising from a low of 43.11 per cent in 1987 to a high of 80.07 per cent in 1992, a large jump from the 1990s 61.15 per cent, but staying at the same level as in 1991.

Improved liquidity due to the issue of shares during the year and the taking out of new loans to repay old ones. Sales of assets exceeded purchases so the group as a whole had an inflow of funds. Though the company is not profitable, liquidity continues to look good and cash conservation is clearly a priority.

Commitments and contingent liabilities are quite heavy and if called in all at the same time could pose problems for the company and the group. Operating lease commitments, while not onerous, could result in severe action for breach of contract if the company were to go into liquidation. Similar actions could be brought for cancellation of the capital expenditure plans already contracted.

4 Inflation

	1988	1989	1990	1991	1992
Rise in turnover	£230M	£240M	£342M	(£266M)	(£39M)
Percentage rise on previous year	10.94	10.29	13.29	(9.10)	(1.45)
Turnover needed to keep pace with inflation	£2,176M	£2,517M	£2,781M	£3,155M	£2,756M
Real percentage rise in turnover	7.22	2.22	4.82	(16.01)	(5.26)
Rise in PBT (BRC)	£45.4M	£19.4M	(£8.5M)	(£154.8M)	(£38M)
Percentage rise on previous year	16.95	16.14	(2.56)	(47.76)	(24.80)
PBT (BRC) needed to keep pace with inflation	£277M	£338M	£360M	£351M	£279M
Real percentage rise in PBT (BRC)	13.07	(1.56)	(9.86)	(51.71)	(59.07)
Rise in DPS	1.10p	1.10p	1.00p	0.00p	(4.50p)
Percentage rise on previous year	15.07	13.10	10.53	0.00	(38.09)
Dividend needed to keep pace with inflation	7.55p	9.06p	10.27p	11.37p	10.92p

Real percentage rise in DPS	11.26	4.86	2.24	(7.65)	(40.48)

Conclusion

Overall, all areas of the business are suffering and real growth in turnover, dividends and profit before tax has been stagnating in previous years and is now falling into the negative. Real returns for shareholders are negative and the share price should decrease in value as a result of this. While the rate of decline on inflation adjusted turnover has slowed dramatically this year, the rate of decline in PBT and dividends continues to grow and gives additional evidence of the severity of the recession that is affecting the group and its profitability.

5 Additional information which would be useful to users

- For pensions, additional information on the pension scheme such as the accounts so that employees can see the true state of the scheme and assess the trustworthiness of the trustees and see where the scheme assets are invested.
- Future budgeted profits and cashflow forecasts.
- Information on competitors.

6 Creative accounting

- Capitalization of interest, though very small this year, with no indication if this is net of tax relief or whether tax relief is taken through the balance sheet and not through the profit and loss account. Therefore possible creativity of tax charge in the accounts.
- Reserve accounting of profit and loss account and other reserves.
- Fair value accounting: assets valued up or down?
- Stock value is down, but this is probably due to falling order books and careful working capital management rather than to creativity.
- Leasing commitments up this year suggesting that sale and lease-back is now becoming a group policy to realize cash and keep assets off the balance sheet.

Resisting the temptation to creativity account:

- no attempt to classify business closure costs as extraordinary
- no attempt to treat profit on sale of fixed asset investments as exceptional, correctly treated as extraordinary
- research and development expenditure is written off as incurred
- compliance with UITF 2 that restructuring costs should be treated as exceptional.

7 User groups

Consider information from the point of view of each user group in terms of:

- Relevance
- Reliability
- Understandability
- Completeness
- Comparability, Objectivity and Timeliness where relevant.

(a) Shareholders: Existing and potential

Concerns: stability, high earnings and dividend potential.

Less than sound company from their point of view with stagnating earnings over the past two years with real growth going into negative and falling dividends. Issues of new shares during the year will dilute their earning power and capital gains prospects. Shareholders may feel some concern about the reduction by the chairman of his holding in the company without further explanation.

(b) Loan creditor group

Concerns: stability, good liquidity and getting paid.

Low liquidity, highly geared and highish debt ratio. Therefore creditors and loan creditors might be a little concerned about the company and group's current position but will be encouraged by the improvement in short term liquidity this year together with the stabilization of gearing and debt ratio.

(c) Employees

Concerns: stability, potential for increased earnings and promotion, coupled with training.

Company is in difficulty, decreasing its workforce (and has always had higher redundancy costs in previous years). Committed to training its employees, though. Pension schemes in force. Only problem with the accounts might be that they are too complex for the financially illiterate.

Share option scheme in place.

Training, employment, pension scheme, staff costs and directors' remuneration note.

Employment policy and practice and employee involvement.

(d) Analyst adviser group

Concerns: openness of company to disclose all information required in format readily understandable and comparable to other companies in the same sector.

Accounts in Companies Act (CA) format, therefore easily comparable to other companies in the sector, some creative accounting but this is generally identifiable.

(e) Business contact group
Concerns: stability, comparability to other trading partners and rivals and continuing ability to trade to provide a long-term trading partner. Company is experiencing difficulties but is a household name and so a good company to trade with in the long term.

(f) Government
Concerns: stability of company to ensure that a continuing flow of taxes and national insurance will be maintained. Interest in the company as an employer of labour in the economy.
Large exporter of goods overseas and large earner of foreign currency.

(g) The Public
Concerns: stability of the company as an employer and as a contributor to the economy as a whole, contribution of the company to the environment as a whole, pollution and its control etc.
Company contributes to the economy and provides general information on its concern for the environment and the community as a whole through charitable donations and arts sponsorship.

Note: Annual reports tend to present a glowing and glossy picture of the company so users should not be misled into thinking that all is totally well.

Appendix 7.2

Case study 7.4: Rolls-Royce

Tough tactics needed to stay in the top three
Restructuring has become a way of life for Mr Terry Harrison, Rolls-Royce chief executive. 'I seem to spend all my time doing it,' he says.

The latest upheaval came yesterday when the UK aero-engine group fell into the red with a 1992 pre-tax loss of £184m, reflecting provisions of £268m; cut its dividend; and announced it would reduce its 51,800-strong workforce by 5,000.

The news was not a total surprise. Rolls-Royce has already shed 12,000 jobs in the past two years. Along with the rest of the aerospace industry, it has been buffeted by what Sir Ralph Robins, chairman, concedes is 'the longest and most damaging recession we've ever suffered'.

Rolls-Royce's two main US competitors, General Electric and Pratt & Whitney, have also announced plans to cut 4,000 and 10,000 jobs respectively over the next 18 months.

'In some parts of the world, we don't know what kind of airline industry will emerge from this recession which has already taken some large airlines like Pan Am and Eastern out of the game', Sir Ralph warns.

The concurrent cut in defence spending following the end of the cold war is also likely to be permanent. 'The defence business has not gone away for good, but what is clear is that it will be half the size it was,' he says.

The turmoil in the aerospace industry has revived memories of 1971 when Rolls-Royce was forced into bankruptcy. The company was dragged down by huge development problems of its RB211 civil aircraft engine. At that time, it was heavily dependent on the aero-engine business, and lacked a broad range of civil aircraft engines.

Sir Ralph is adamant that the company will never repeat the mistakes of the past. It launched a recovery strategy eight years ago to improve its aero-engine model range, reduce costs, improve efficiency and broaden its industry base. 'We believe this strategy continues to be the right one,' says Sir Ralph. 'The fact there is short-term turbulence in the market should not divert us from our long-term aims.'

The strategy has started to pay off. Restructuring has led to a 13 per cent improvement in output per employee. Continuing rational-ization will close some sites, including the helicopter engine plant at Leavesden in Hertfordshire and the aero-engine component facility at Parkside, Coventry. At the same time, the company is consolidating some smaller facilities into larger ones.

But the job cuts will not go on for ever. 'We want to become more efficient but we also want to maintain our production capability to take advantage of the recovery when it occurs,' Sir Ralph empha-sized.

In the industrial power activities, the workforce has already fallen from 27,000 in 1990 to 22,300 at the end of last year, and is likely to remain stable. In the aerospace division, it has dropped from 36,500 in 1990 to 29,500 at the end of last year, and the company's eventual target is around 25,000.

The company has also broadened its civil engine range. Its engines now cover about 80 per cent of the commercial aircraft market compared with barely 30 per cent a decade ago.

Even in a depressed market, Rolls-Royce is selling more engines than before. 'Until 1989, we were selling about 100–150 engines a year; it's now around 4000 a year because we can offer products to power many more different airliners than in the past,' says Sir Ralph. In the longer term, this will lead to a steep increase in the company's engine spare parts business, a traditional source of high margin business.

Rolls-Royce has also seen its share of the civil engine market increase. Sir Ralph said the company had captured about 22–23 per cent of the civil aero-engine market compared with 26–28 per cent each for GE and Pratt & Whitney.

Equally important has been the expansion of the company's industrial power operations, initially through the acquisition of Northern Engineering Industries (NEI) in 1989 and then through the alliance with the US Westinghouse group eight months ago.

The industrial power group already accounts for about 40 per cent of the company's £3.5bn annual turnover. Rolls-Royce wants these activities, which last year helped offset the slump in the company's aero-engine business, to grow eventually to around 50 per cent of turnover. This could involve more joint ventures and possible acquisitions, explains Mr Harrison, former head of NEI.

The other big challenge facing Rolls-Royce is the £400m development of the Trent, its new heavy thrust civil engine to power the next generation of large wide-body airliners. With these new big jets expected to account for a growing share of the civil aircraft market over the next twenty years, Rolls-Royce cannot afford to slow down on the development of the Trent which is absorbing about half the company's total annual £200m research and development expenditure. The competition for this new market is already intense. Roll-Royce's two US rivals are also developing large engines and some aerospace analysts believe that three competing engines are too many. Some argue that three big aero-engine manufacturers are also too many and one will eventually have to merge with a competitor.

'I think the three majors will survive but I can't see a place for the smaller players,' says Sir Ralph, adding that the next few years are likely to see the consolidation of smaller companies into the big three.

Sir Ralph also believes that the short-term problems facing the three main aero-engine manufacturers are greater than the longer term. 'I don't see a change in the world's desire to travel by airplane: the opportunity for us is clearly still there,' he says.

He concedes that when the market does ultimately pick up, it is unlikely to roar ahead as in previous cycles. Instead, it will probably take the industry along a slow, scenic route. In the meantime, there will be no escape from restructuring. 'This business is not for the faint-hearted,' Mr Harrison says.

Financial Times 12 March 1993

Group financial highlights for the year ended December 31, 1990

	1990 £m	1989 £m
Turnover	3,670	2,962
Operating profit	468	383
Research and development (net)	(237)	(161)

Profit before exceptional items and taxation	226	237
Profit before taxation	176	233
Profit attributable to the shareholders	134	192
Shareholders' funds	1,164	1,126
Earnings per ordinary share – net basis before exceptional items	19.1p	21.8p
Dividends per ordinary share	7.25p	7.0p
Average number of ordinary shares in issue	961m	901m

Summary group profit and loss account for the year ended 31 December 1990

	1990 £m	1989 £m
Turnover	3,670	2,962
Operating profit	468	383
Research and development (net)	(237)	(161)
Income from interests in associated undertakings	2	–
Net interest (payable) receivable	(7)	15
Profit on ordinary activities before exceptional items and taxation	226	237
Exceptional items*	(50)	(4)
Profit on ordinary activities before taxation	176	233
Taxation	(36)	(36)
Profit on ordinary activities after taxation	140	197
Attributable to minority interests in subsidiary undertakings	(6)	(5)
Profit attributable to the shareholders of Rolls-Royce plc	134	192
Dividends – interim paid 2.55p (1989 2.3p) per share	(24)	(22)
– final proposed 4.7p (1989 4.7p) per share	(45)	(45)
Retained profit for the year	65	125
Earnings per ordinary share		
Net basis	13.9p	21.3p
Net basis before exceptional items	19.1p	21.8p

*A provision of £50m has been made to cover restructuring costs and to provide for uncertainties faced by customer airlines.

Analysis by business segment	Turnover		Profit	
	1990 £m	1989 £m	1990 £m	1989 £m
Aero gas turbines**	2,340	2,054	74	147
Power engineering	976	674	82	52
General engineering	354	234	27	19
	3,670	2,962	183	218

The profit represents 'Profit on ordinary activities before taxation' as adjusted for net interest (payable) receivable.
**The exceptional items of £50m (1989 £4m) have been charged against Aero Gas Turbines results.

Summary group balance sheet at 31 December 1990

	1990 £m	1989 £m
Fixed assets		
Tangible assets	676	658
Investments	36	25
	712	683
Current assets		
Stocks	888	754
Debtors	816	749
Short-term deposits and cash	431	407
	2,135	1,910
Creditors – amounts falling due within one year		
Borrowings	(100)	(52)
Other creditors	(1,074)	(939)
Net current assets	961	919
Total assets less current liabilities	1,673	1,602
Creditors – amount falling due after one year		
Borrowings	(161)	(162)
Other creditors	(127)	(68)
Provisions for liabilities and charges	(183)	(173)
	1,202	1,199
Capital and reserves		
Called up share capital	192	192
Share premium account	239	238
Revaluation reserve	132	135
Other reserves	22	28
Profit and loss account	579	533
Shareholders' funds	1,164	1,126
Minority interests in subsidiary undertakings	38	73
	1,202	1,199

Group five year review
for the year ended 31 December

Profit and loss account

	1990 £m	1989 £m	1988 £m	1987 £m	1986 £m
Turnover	3,670	2,962	1,973	2,059	1,802
Operating profit	468	383	333	354	276
Research and development (net)	(237)	(161)	(149)	(187)	(132)
Income from interests in associated undertakings	2	–	–	–	–
Net interest (payable) receivable	(7)	15	13	(4)	(21)
Profit before exceptional items and taxation	226	237	197	163	123
Exceptional items*	(50)	(4)	(29)	(7)	(3)
Profit on ordinary activities before taxation	176	233	168	156	120
Taxation	(36)	(36)	(22)	(21)	1
Profit on ordinary activities after taxation	140	197	146	135	121
Attributable to minority interests in subsidiary undertakings	(6)	(5)	(1)	(1)	(1)
Profit attributable to the shareholders	134	192	145	134	120
Dividends	(69)	(67)	(50)	(42)	–
Retained profit for the year	65	125	95	92	120
Earnings per ordinary share					
Net basis	13.9p	21.3p	18.1p	18.2p	18.9p
Net basis before exceptional items	19.1p	21.8p	21.7p	19.1p	19.4p
Dividends per ordinary share	7.25p	7.0p	6.3p	5.25p	–

Balance sheet

Fixed assets	712	683	473	438	405
Current assets	2,135	1,910	1,309	1,106	940
	2,847	2,593	1,782	1,544	1,345
Liabilities & provisions	(1,645)	(1,394)	(830)	685)	(833)
	1,202	1,199	952	859	512
Share capital	192	192	160	160	127
Reserves	972	934	789	695	380
Shareholders' funds	1,164	1,126	949	855	507
Minority interests in subsidiary undertakings	38	73	3	4	5
	1,202	1,199	952	859	512

Other financial information

Research and development (gross)	480	343	304	328	299

*In 1990 an exceptional provision of £50m has been made to cover restructuring costs and to provide for uncertainties faced by customer airlines.

8 Management accounting

In Chapters 6 and 7 we examined Phase 1 of the accounting process (t_{-1} to t_0), which uses records of financial transactions to generate summarized historical financial reports information for users essentially external to the organization; shareholders, lenders and investors for example. In this chapter, we examine Phase 2 of the accounting process (t_0 to t_1) where different information is generated for managers in engineering, sales and production.

This does not mean the user group information sets are mutually exclusive. It is rather that managers at all levels would find the information contained in a set of financial accounts as being of insufficient detail to assist in specific decision making tasks and carrying out operational responsibilities. On the other hand, it could be argued, this internal information is too detailed to enable external users to build up an accurate overview of the company's financial position, and there would be considerable peril in making such information available to competitors.

The engineering manager, along with other managers, needs to use management accounting to assist in planning, controlling and decision-making. Management accounting can be regarded as providing relevant, accurate and timely information concerning the acquisition and deployment of the economic resources used by the organization to meet its objectives. In particular, it focuses on the identification, measurement and

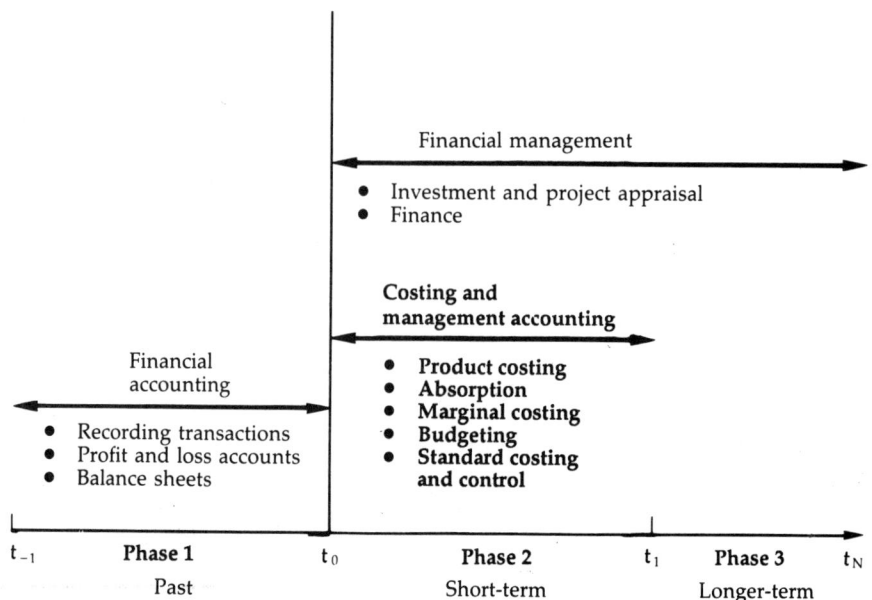

Figure 8.1 Phase 2: time horizons and management accounting

Financial management
- Investment and project appraisal
- Finance

Costing and management accounting
- **Product costing**
- **Absorption**
- **Marginal costing**
- **Budgeting**
- **Standard costing and control**

Financial accounting
- Recording transactions
- Profit and loss accounts
- Balance sheets

t_{-1} **Phase 1** t_0 **Phase 2** t_1 **Phase 3** t_N
  Past   Short-term   Longer-term

control of costs. Since engineers make a major contribution to the competitiveness and profitability of the organization by driving down production and design costs, there is clearly a vital overlap here.

Unlike financial accounting, there are no universally set procedures, preordained legal or quasi-legal formats that have to be adhered to. In fact, it is entirely optional for management accounting information to be provided at all.

Since management accountants are primarily concerned with costs, we start by examining the major costing methods used by industrialists.

Cost accounting

There are two fundamental methods for costing a project, product or service. They are:

1 Absorption (or Full) costing
2 Marginal costing.

Absorption costing is the method recommended for assisting the financial accounting process in undertaking a score-keeping role for the preparation of financial accounts, as referred to earlier. The idea is that all the firm's costs are covered by the sum of its profitable activities, by working out the cost of each unit of output made and recovering costs through appropriate pricing.

Marginal costing, on the other hand, examines individual projects in isolation from the rest of the firm's activities. It works by attempting to model the effects on costs of management decision-making in the short-term. As with absorption costing, the process ends with an estimate of costs per unit; but marginal costing takes into account only the extra costs incurred by increasing output by one unit.

Case study 8.1: Alex's Nuts

Alex runs a pub. A savoury snack representative has suggested introducing a new line – peanuts sold from a dispensing machine. The machines costs £250 and the peanuts will sell for 25 pence. The purchase price of the peanuts is 15 pence. Alex naturally assumes that he will make 10 pence per packet. If 100 packets per week are sold he will make £10 profit. After 25 weeks of the same sales he will have paid for the dispensing machine and thereafter make 10 pence pure profit per sale. The machine is duly installed.

Bill Beanie, a regular customer, is an accountant and couldn't help notice the machine. Alex explained his business logic to Bill. 'Very foolish – you have been had, Fred,' said Bill. 'How?' enquired a peeved Alex. 'Well,' said Bill, 'you have to pay for the upkeep of this pub and pay your bar staff. This peanut operation should reflect some of the costs of this – a conservative estimate would suggest at least £600 per annum would be a fair proportion of the total. Also, you have to pay for the upkeep of the place with cleaning and tidying, which must another £200 per annum. By my reckoning if £800 per annum is allocated for these overheads; the costs work out at 16 pence per bag. When you add the purchase price of 15 pence it means you are making a loss of 6 pence before you start – at that rate you will never pay for the dispensing machine. Take it from me, if you really want to make a profit of 10 pence per packet you should be charging at least 41 pence.' 'Nobody will buy them at that price!' protested Alex, and promptly disposed of the dispensing machine.

Alex's original thinking reflects a marginal cost approach. Bill Beanie has examined the same activity from an absorption cost approach.

Since both methods provide a means of estimating cost per unit, it is worth reflecting on why Alex should need to calculate cost per unit anyway. Here, we show the importance of costing on the profit figure calculated in the PLA, and the impact of costing on the pricing decision.

Impact of costing techniques on the pricing decision

1 Profit measurement

In a multi-product organization, the way in which indirect costs, such as administration, cleaning and maintenance of premises, sales engineer's expenses etc. are added to the direct cost (such as labour and materials), of producing a particular product, will have a dramatic effect on the periodic costs measured, and therefore profit.

Furthermore, in determining gross profit, the cost of sales has to be subtracted from the sales revenue figure. The cost of sales in turn, under the accruals concept of accounting, can only be determined after the closing stocks in the business have been valued at the end of the accounting period. These closing stock values will then be carried forward into the next period. Each stock value must include a proportion of costs reasonably incurred in getting the product into its finished state, ready for sale. These costs will include indirect costs that do not really lend themselves to being directly measured relation to the output, like the cleaning costs in Alex's bar, but are nevertheless incurred because output takes place. Again, the way in which these costs are allocated to individual products will influence the profit figure.

2 Product costing

In Chapter 2 we pointed out that many factors influence the price charged to the consumer. However, the cost of producing the product is clearly of vital importance and the price charged must cover these costs; in the long term at least. Clearly, it is risky (but by no means unknown) to price products when appropriate cost information is not known.

A firm producing or providing one standardized product or service will find it difficult to survive if prices are less than costs for anything other than a very short space of time. There may be valid short-term marketing objectives that involve pricing at less than costs, but these would need to be closely monitored as to their effectiveness.

In the non-standard product firm, it is critical that the organization can determine what the costs of the product or service will be, because the price will be specified in a written contract agreed in advanced. If the contract is relatively large then underestimating costs and prices could have serious consequences for the organization.

In a multi-product firm, some products can be priced at less than the full cost as long as others are more than recovering the costs allocated to them. The reasons for this are numerous: spare capacity, tactical marketing such as loss-leading or a new market entry. But these are issues that lend themselves to decision-making and marginal costing, examined later in this chapter. In the longer run however, it is rare for a product to be sustained if costs cannot be recovered through prices.

Absorption costing

If an organization provided one type of standardized product throughout the year, the cost of making one unit could easily be determined through the existing financial accounting system. It would simply mean dividing the total costs of the organization as per annual Profit and Loss Account and dividing by the output produced for that year. There would be little need for any more sophisticated system.

But as firms grow and reach a particular size, then the range of products and/or services is likely to increase. Further, many organizations in the engineering sector do not provide standard products, and have to work to bespoke customer specifications. For example, a civil engineering contractor may specialize in bridge-building but each bridge will be designed individually, meeting different technical, geological, economic and design criteria.

Absorption costing is a method adopted to spread all of the costs of the organization equitably over its different outputs over a period of time. This means trying to determine the cost of each job, contract or standard product made. The basis of absorption is to identify the production costs first. This is a conventional approach that does not accommodate the greater prominence given to marketing, distribution and sales activities over recent years i.e. the total product offering, not simply the physical product.

The process of absorption costing will be dealt with under the following stages: first, traditional costing terminology; second, how costs are accumulated internally and allocated; third, once costs have been allocated how they are then apportioned; and finally, how costs are absorbed into the final output. Figure 8.2 shows the absorption process outlined here.

Absorption costing terminology

Much of the terminology has developed from the days when organizations were manufacturing orientated i.e. whatever was produced was sold. Costs over and above manufacturing were likely to involve only selling and administrative costs.

Under this approach there are three broad cost categories: materials, labour and expenses (i.e., costs other than material and labour).

All would be incurred within a production environment. This categorization is then divided into two sub-groups: direct and indirect costs.

Direct costs:

These are costs that can be traced (directly) to the product. For example, the physical materials used in the manufacture and presentation of the product to the customer are readily measurable and identifiable. Labour can be traced to a product if the use of that labour is devoted exclusively

Figure 8.2 Absorption
costing process

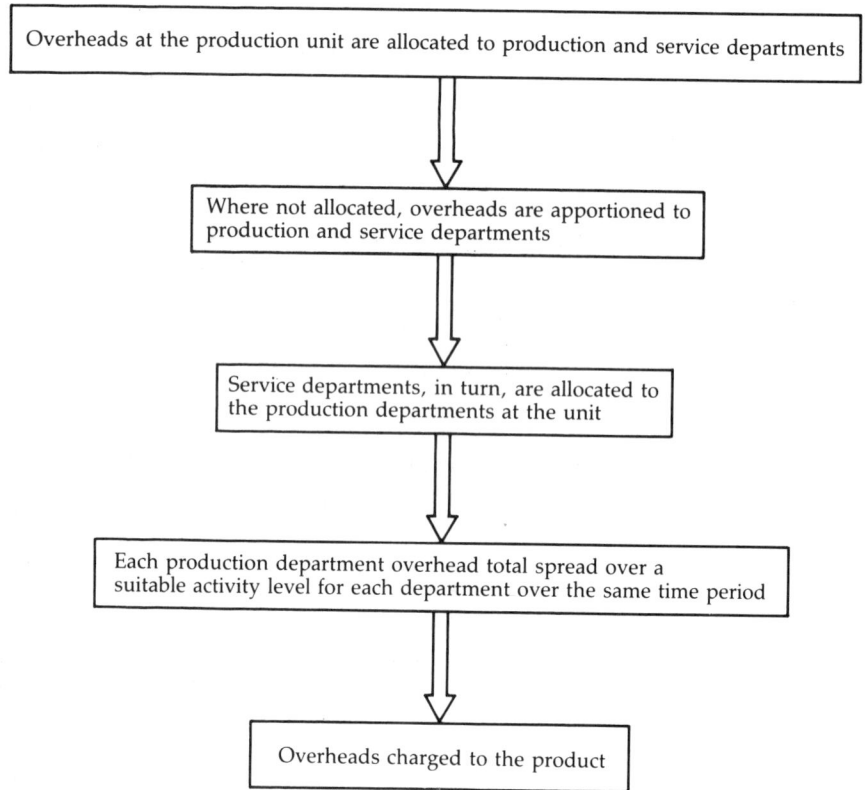

```
┌─────────────────────────────────────────────────────────────────────┐
│ Overheads at the production unit are allocated to production and service departments │
└─────────────────────────────────────────────────────────────────────┘
                                      │
                                      ▼
          ┌─────────────────────────────────────────────────┐
          │ Where not allocated, overheads are apportioned to │
          │ production and service departments                │
          └─────────────────────────────────────────────────┘
                                      │
                                      ▼
            ┌─────────────────────────────────────────────┐
            │ Service departments, in turn, are allocated to │
            │ the production departments at the unit         │
            └─────────────────────────────────────────────┘
                                      │
                                      ▼
       ┌──────────────────────────────────────────────────────────┐
       │ Each production department overhead total spread over a    │
       │ suitable activity level for each department over the same time period │
       └──────────────────────────────────────────────────────────┘
                                      │
                                      ▼
                ┌──────────────────────────────────────┐
                │ Overheads charged to the product       │
                └──────────────────────────────────────┘
```

to that product. The categories of expenses directly traceable to products depends on the nature of the business. If a machine was devoted solely to producing a standard product over a period of time, then the costs associated with that machine would be considered direct costs; such as depreciation, cleaning, and energy. Royalty and licence costs could be considered as direct. Below, we look at materials and labour in greater detail.

Indirect costs

Indirect costs are alternatively known as overheads. They use the same three cost categories as direct costs, but they cannot be directly traceable to the product economically. They would include, for example, general supervisory services on the production line (although if supervisory labour were allocated to a particular product then it could be classed as direct).

This classification gives the impression that there are certain costs that are always direct or indirect. Although this may be so, the real criteria is one of materiality. If the cost of providing the information to attribute directly a cost to individual products is greater than the benefits of providing it, it would be uneconomic to trace the cost and therefore it would be treated as indirect. For example, energy costs may not be broken down by individual products unless an electronic metering system is installed. Without the system, the energy costs would be indirect, but with the system installed the costs become direct. A firm with such a meter will treat the costs differently to a firm without.

Allocation of costs

Direct materials

In Chapter 6, consideration was given to the acquisition of materials, their storage, conversion into WIP and finally into finished product ready for sale.

The purchase of materials and their conversion into final output means that until the product is sold (or materials used up during the engineering and production process) they are assets of the business. While they are assets they need to be carefully tracked by stock control procedures.

A major part of materials cost will be the price originally paid. From an accounting perspective, these prices need to be recorded systematically, because they will be traced to the overall cost of the output. This is complicated by the tendency of the current market price to rise (or occasionally fall) while the item is held in stock. The reasons for this are many, inflation, sudden exchange rate changes, poor or good harvests and a change of supplier may all produce this effect. Storage of materials means that identical raw material stock is distinguished only by it being purchased at a different price. From an internal costing point of view there are a variety of prices that could be charged, and as a consequence, a variety of differing costs in the same accounting period. We have briefly mentioned FIFO (first in, first out), in Chapter 6. Here, we examine the issue more fully.

For example, the prices paid for component Z over a two month period, March and April, was £12 and £14 per unit respectively. Each purchase was a batch of 100 units. In May the production department used 160 units.

Month	Units	Input Price £/unit £	Value £	Output Units	Cost £	Stock Stock units	Stock value
February	0	12	0	0	0	0	0
March	100	12	1200	–	–	100	1200
April	100	14	1400	–	–	200	2600
May	0	0	0	160	?	40	?

The '?' reflects the differing approaches acceptable for costs of output and periodic stock valuation. There are three basic approaches: FIFO, LIFO and a weighted average cost.

Under first in first out (FIFO) principles, the oldest stock is charged to the cost at the old price. This will mean that stock is valued at the most recent price of purchase for balance sheet purposes.

Under last in first out (LIFO), the most recent stock is charged to cost first. Stock at the old prices is valued in the balance sheet.

The weighted average (AVCO) is a hybrid approach between FIFO and LIFO. The question marks will be filled in according to the values generated by each convention in turn.

FIFO

Cost of stock used: 100 units @ £12 = £1200
60 units @ £14 = £840

£2040 = cost to be charged to WIP finished goods or cost of sales

Value of remaining stock: 40 units @ £14 = £560 = stock valuation end of May for balance sheet

£2600 = purchase price of input

LIFO

Cost of stock used: 60 units @ £12 = £720
100 units @ £14 = £1400

£2120 = cost to be charged to WIP, finished goods or cost of sales if sold

Value of remaining stock: 40 units @ £12 = £480 = stock evaluation for end of May balance sheet

£2600 = purchase price of input

AVCO

Cost of stock used: 100 units @ £12 = £1200
100 units @ £14 = £1400

Value of remaining stock: 200 units @ £13 = £2600
less
160 units @ £13 = £2080 = cost to be charged to WIP finished goods or cost of sales

40 units @ £13 = £520 = stock valuation end of May for balance sheet

£2600 = purchase price of input

All three methods are internal mechanisms for determining how outputs should be charged in a period and how the stocks would be valued. In all cases the documented purchase of £2600 does not alter – only the allocation between expenses for the current profit measurement period and assets for the next period.

Generally, when procurement prices are increasing then:

cost of sales   LIFO > AVCO > FIFO
thus closing stock LIFO < AVCO < FIFO
therefore profit   LIFO < AVCO < FIFO

Therefore, the periodic reporting of profits can differ, all other things being equal, because of the differing methods of stock valuation being applied. For external reporting purposes both FIFO and AVCO are acceptable methods – LIFO isn't. Whichever method is selected, the consistency convention outlined in Appendix 6.1 implies that the same technique should always be used year on year, regardless of which method suggests a more attractive picture in any one year. For internal management purposes, any of the methods are acceptable. LIFO may provide a more realistic profit measure in inflationary conditions. Indeed there are other methods such as replacement value (or, NIFO – next in first out) that may give an even more conservative measure of profits when prices are rising – but again this is not valid for external reporting.

For computerized stock records, the AVCO approach is often taken, because profit reporting will generally be more conservative than FIFO if the climate is inflationary.

Direct labour

These are wages and salaries paid to employees directly concerned with the product manufacture or service provision. Examples include: machinists, assembly operatives, bricklayers etc. There are a number of methods for determining direct labour costs in the overall product. Most of the information can be derived from the payroll information outputs. There are broadly two systems of paying employees, time-based, and output based. 'Clock-cards' or 'time-sheets' which indicate the hours or days worked at a specified rate per hour or day will consistute part or all of the total direct labour cost. 'Piece-rate' schemes effectively only pay upon a rate per unit of measured output or operation. Incentive schemes are a combination of both methods whereby a minimum amount of earnings is guaranteed, usually on the amount of time worked, together with additional earnings for achieving higher output levels than planned or productivity improvements. Most engineering employees are paid salaries and therefore labour costs tend not to vary with output. More recently profit-related pay and bonuses are becoming part of the remuneration package. Thus it is becoming more difficult to directly attribute labour costs from traditional methods.

Direct expenses

Any cost or expense other than labour and materials that can be directly traceable to the product should be identified. These types of costs are usually difficult to trace, or uneconomic to trace, and are therefore categorized as indirects or overheads (see indirects). However royalties, patents and licence agreements are costs that should be traced to the particular output. For example, the development of a commercial software package may be reliant on purchasing some ready developed copyrighted software (such as Windows or Lotus 1–2–3) from other companies as an input and thus, every software package produced would involve a royalty or licence payment to the providers of the software.

Once the three categories of direct cost have been identified and measured, they are summed to produce the prime cost of production:

Direct materials
+ direct labour
+ direct expenses

= Prime cost of production

The prime cost is that part of the cost of the product where costs can be directly traced to the output. Japanese style production methods, together with increases in technology enable more costs to be traced direct to the product. Thereafter, the product cost is determined by the apportionment of pooled overhead costs to the product. From a managerial perspective, the prime cost can be a useful measure to compare over time for control purposes, and is shown in the manufacturing profit and loss account.

Indirect costs

Some indirect costs of production can be directly attributed and allocated to individual products and production runs. They would normally be categorized in the same way as direct costs; that is,

- Indirect materials: such as cleaning fluids and lubricants
- Indirect labour: such as machine shop supervision
- Indirect expenses: such as heating and lighting in the machine shop

These would them be summed to give a cost for factory overheads:

$$
\begin{array}{l}
\text{Indirect materials} \\
+ \text{ indirect labour} \\
+ \text{ indirect expenses} \\
\hline
= \text{Production overheads}
\end{array}
$$

The cost of production is simply:

Prime cost + Production overheads

This total would then be divided by the output to give an idea of the cost of actually producing the product. This is the first stage of absorption, that is, allocating production costs to individual products.

However, this total, although extremely important, does not account for the total costs incurred by the firm. There are other overheads that are not easily attributable to particular products. These costs are usually grouped together in functional areas such as sales, research and development, marketing and administration. These other costs must be somehow incorporated into the calculation to reflect the cost per unit of the organization's output. The means by which these disparate activities are somehow tied to the production process is through the identification of cost centres, and the allocation of the costs of each cost centre to productive activities.

The cost centre

'A location, function or items of equipment in respect of which costs may be ascertained and related to cost units for control purposes' (CIMA). For example, a laboratory would be a cost centre if it produced no output for sale outside the organization. Any enquiry from other departments in the organization would be charged to that department through a cost centre code system.

The organization is thus divided up into administrative pigeon-holes that enable costs to be allocated to particular cost centres as they occur. These 'pigeon-holes' are expressed in numerical codes. The cost centre code is thus used as an administrative base so costs can readily be identified.

In the production unit there will be a number of cost centre codings. For each stage of production (which could be departmental) there would be a separate code. In addition, for each supporting service department there would be a unique code.

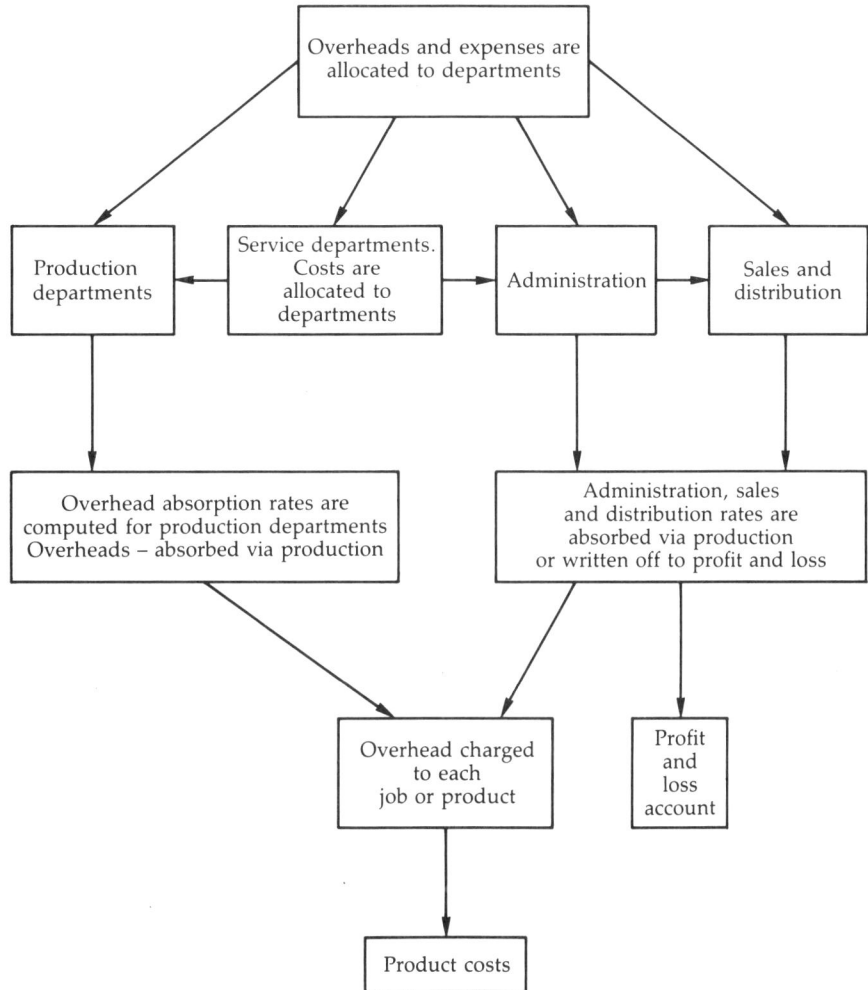

Figure 8.3 Allocating costs to cost centre *Source*: Chadwick and Magin.

Codes could be specified and listed on a computer and any official documentation relating to costs could be categorized and input accordingly. The costs could be automatically stored, allocated and apportioned by computer software.

Apportionment of costs

It is important that the appropriate costs of the organization are traced to the appropriate cost centre. If the engineering services department purchases materials then they are charged to the cost centre coding allocated to the department.

Costs are not so easy to allocate to the engineering services department if they are incurred for the benefit of more than one cost centre. Suppose for example, the rent incurred for production facilities in 199X was £20,000. For external reporting purposes the cost to the firm is £20,000 and would simply be regarded as a transaction for financial accounting purposes. However, internally, there is a need to identify the use of resources to provide costing information.

If the £20,000 were to benefit the manufacturing and distribution cost centres besides the engineering services department, then a rational basis for apportionment is needed. This might require seeking additional non-financial information to enable an equitable apportionment of those costs. If rent was charged to the company on the basis of square metres occupied, then logic might suggest that the rent is split *pro rata* to each cost centre on the square metres occupied.

Activity

How much rent should be charged to each cost centre if the measured occupancy of the premises totalled 40,000 sq.m. and the manufacturing, distribution and engineering services occupied 25,000, 10,000 and 5,000 sq. m. respectively?

Secondary apportionment

The costs of a product are collected in their respective production cost centres. In addition, there will be productive support departments that are not directly producing but indirectly facilitating the production departments. These are regarded as service departments, and would include maintenance departments, work-study etc.

For example, maintenance departments will often be required to service all the productive areas of the business. Thus for a textile firm involved in cutting, dyeing, machining, finishing and packaging, the maintenance costs would need to be spread as equitably as possible over the individual production departments. Because the activity of maintenance in this case is labour intensive, an equitable basis for the apportionment could be the labour time logged for maintaining each productive cost centre over the time period.

In arriving at the overall cost of the product it might not appear to be necessary to allocate the maintenance service department costs to each productive cost centre. To some extent this would be true if the unit output was sequential and relied upon the finished goods only being completed by going through all five departments. However, valuable management information could be lost because an analysis could be undertaken on the maintenance department. If it was found that 70 per cent of maintenance costs were being spent in the packaging department, this could be monitored and investigated if needed.

Case study 8.2: Grinders Mill

Grinders Mill is a traditional engineering firm. It has two main production lines, organized into separate departments (Production 1 and 2). In this example, we look at the way the costs of two service departments, Engineering Services and Design, are allocated to the two production departments.

All overheads have been identified and allocated and apportioned throughout the business. The production and engineering overheads for the year were as follows:

Cost centre	Production and engineering cost centres				
	£	£	£	£	£
	Produc-tion 1	Produc-tion 2	Engin-eering service	Engin-eering design	Total
Overheads: Allocated and apportioned	11,000	9,000	10,000	8,000	38,000

Additional information

The activity of the engineering services has been measured in terms of requisitions from the other departments. The engineering design was analysed according to labour-hours per design. A breakdown of the requisitions and labour-hours was undertaken and expressed in the following percentage terms:

Cost centre	Production 1	Production 2	Engineering service	Engineering design
Engineering services	30%	50%	–	20%
Engineering design	70%	30%	–	–

Worked solution

To date, all four cost centres – two production and two service – have had indirect costs allocated and apportioned. However, the service cost centre costs have to be charged to the production cost centres in proportion to the use of their services.

If the above proportions are used for each department's costs then:

	Produc-tion 1	Produc-tion 2	Engin-eering service	Engin-eering design	Total
	£	£	£	£	£
Indirect costs	11,000	9,000	10,000	8,000	38,000
Engineering	3,000	5,000	(10,000)	2,000	–
services: (30:50:20)	14,000	14,000	–	10,000	38,000

Thus all the costs of Engineering services have been apportioned to other cost centres. The same process is used to allocate the costs of Engineering design.

Engineering design: (70:30)	7,000	3,000	–	(10,000)	–
Overheads	21,000	17,000	–	–	38,000

> At this stage, Engineering services and Engineering design have been apportioned on an equitable basis.
>
> This now means that each production cost centre can spread their total overhead costs over the productive activity of that production cost centre in the form of a ratio called the overhead absorption rate (OAR). That is, the sums £21,000 and £17,000 will be divided up to give a cost per unit produced, per machine hours used or per labour hour used as appropriate.

The overhead absorption rate

The expressed output can be in the form of units if the production activities are identical units. More commonly, output is expressed in a further denomination of activity such as machine hours or labour hours.

The general rule is to select that measure of output that causes the majority of the overheads to be incurred. In other words, there is no purpose in expressing overheads caused primarily by machinery if the incidence of those same overheads is caused by the employment of labour such as supervision costs, training costs, canteen costs etc.

Activity

On what basis should the costs of an engineering design department be spread?

Once all the indirect costs of production have been allocated, apportioned and secondary apportionment has taken place then, then all the overhead costs will now be attached to the production cost centres.

Where products are bespoke according to customer specifications such as machine tools, printing, software development etc. then, the direct materials and labour costs can be attached to the particular job or contract. However, under absorption costing, it is important that a proportion of overheads is picked up as the product begins to take shape.

If in our Grinders Mill example, Production 1 had five machines operating for 50 weeks at 42 hours per week then, if the incidence of overheads was primarily because of the use of machines, the activity level would be expressed as:

5 machines $\times$ 50 weeks $\times$ 42 hours = 10,500 machine hours

The absorption rate would therefore be expressed as:

$$\frac{\text{overheads over the time period for production centre}}{\text{activity level for the same time period.}}$$

Using our example the overhead absorption rate (OAR):

$$\frac{£21,000}{10,500 \text{ machine hours}} = £2 \text{ per machine hour}$$

In other words, for every hour of activity the machine is used in producing a product, there will be a charge of £2. If a product used 100 hours of that machine time, it would be charged with $100 \times £2 = £200$.

Activity

Assume Production centre 2 was mainly labour intensive. There were eight staff employed to work fifty weeks at 42.5 hours per week. Determine the OAR.

(Answer: £17,000 / (50 × 42.5 × 8) = £1 per labour hour)

It must be recognized that the above example is simplistic for two reasons: first, manufacturing and engineering processes are more numerous and will require more than the four cost centres identified, and; second, there is the matter of service cost centres benefiting one another. Nevertheless, the principles remain the same but the application will need the help of greater recording and computation facilities.

Also the example referred to the historical recording of overheads and activity. In practice, there needs to be a continuous process of charging overheads to products or services as the activities are being undertaken – not by charging overheads retrospectively. This is particularly important for engineering cost estimating and pricing purposes where potential customers are unlikely to wait until the period end. Therefore, the absorption of the overheads is undertaken in advance of them being incurred by using a predetermined rate (see budgeting in Chapter 9):

Predetermined (budgeted) OAR =

$$\frac{\text{Budgeted overheads for productive cost centre}}{\text{Budgeted level of activity over the same period}}$$

By smoothing the occurrence of the overheads through the use of a rate avoids the problems of fluctuations in the timing of overheads on to the product cost. The swings in costs from one period to another are thus nullified, giving more confidence to managers and customers both.

From a financial accounting perspective there would need to be a retrospective reconciliation between the actual overheads incurred and the actual level of activity. To avoid any major adjustments, it is necessary to ensure proper attention is given to the forecasting of the predetermined rate.

If a budgeted OAR is used for absorbing overheads to output (units), or activity (machine hours etc.) then, at the end of a period, there will need to be a reconciliation between the actual activity and actual overhead costs with the budgeted OAR for correct financial recording of historic profit measurement.

If, for example, budgeted overheads at Grinders Mill were £200,000 and the budgeted level of activity was 100,000 machine hours at the beginning of the year, then the budgeted level of activity would be £2 per machine hour. If, at the end of a measurement period, the actual level of activity was only 80,000 machine hours, then the costs recovered would be only £160,000 (80,000 × £2 per machine hour worked). In addition, should the actual overheads be £240,000, there would be a further discrepancy. There would have been an under-absorption of overheads comprising of two elements:

1 £160,00 – £200,000 = £40,000 due to lesser activity than budgeted, and

2 £200,00 – £240,00 = £40,000 due to greater actual overheads than
      budgeted
     ——————————
     £80,000 under-absorption
     ——————————

This under-absorption would be added to the financial cost of sales to reduce profits for external reporting purposes.

The former is known as a volume variance and the latter an expenditure variance on overheads. Of course with the benefit of hindsight, a budgeted OAR of £3 (where £240,000/80,000 machine hours) would have avoided the problem entirely.

The reverse of this process also holds and there would be an over-absorption and profits would increase accordingly.

Product costing

At this point it is possible to use these calculations to estimate the cost of a product or job – which is the main point of the costing activity. The way this is done will depend on the type of production undertaken. In this section we look at job costing, batch costing and contract costing.

Job costing

Job costs are derived by attaching costs to a particular customer specified product or service. Usually the job is given a particular coding with detailed technical specifications. Jobbing is very common in the engineering industry, particularly in design and manufacture of machine tools, boiler fabrications, design etc. Each job would have its own bespoke customer specifications. The same process of cost build-up would be applied starting with prime costs and absorbed overheads. The job cost can be monitored against the original estimate to enhance management control. Also, the job cost will enable the profitability of the job to be assessed after completion. Any incomplete job will be considered as WIP for stock valuation purposes. The following format is a type followed for job costing.

Job specification XEZ:

	£	£
Direct materials:		
Component W	X	
Component X	X	
	——	
		X
Direct labour:		
Grade 2	X	
Grade 4	X	
	——	
		X
Direct expenses:		
Licence fee		X
Prime cost		X
Overheads absorbed		X
Production cost of completed job		XX

The costing approach taken by the manufacturing organization will attempt to measure the resources used as the product (or service) is being transformed into its final state (or service undertaken). The job cost will conform to customer specifications in detail. Direct costs such as materials and labour are charged to the job cost. In addition the overheads are absorbed on the basis of activity determined by that particular business in each production stage where absorption takes place.

In addition to bespoke engineering firms, many other businesses and professions use this method, or some variation of it, including accountants, solicitors, architects and civil engineers. For estimating purposes in the engineering sector, it is important to ensure that any anticipated recovery of costs is achieved before price quotations are given to customers. Once prices have been agreed and contracts drawn up, it is important to control the costs that have been estimated. Any inefficiencies in output between estimation and actual will have to be carried by the firm – not the customer.

Using the Grinders Mill example above, there are two production departments and two absorption rates: £2 per machine hour and £1 per labour hour respectively. A customer enquiry meant producing a bespoke engineering product using both production departments with the following data:

Production department 1

Component A:	1 @ £4 per unit
Material X:	2 metres @ £2.50/metre
Labour:	4 hours @ £4/hour
Machine time:	5 hours

Production department 2

Material Y:	1 unit @ £5 per unit
Labour:	6 hours @ £3 per hour
Machine time:	1 hour

The total cost of the product would be, using the above data:

Materials:

	£	£	£	£
Component A	4			
Material X	5			
Material Y	5			
		14		
Labour:				
Prod 1 (4 × £4)	16			
Prod 2 (6 × £3)	18			
		34		
Direct costs			48	
Production overheads absorbed				
Prod 1 (5 × £2)	10			
Prod 2 (6 × £1)	6			
		16		
Total production cost			64	

In Production department 1, the incidence of overheads is caused mainly by the use of machines. If, in meeting the customer specification, it was estimated that five machine hours were involved then £10 would be automatically absorbed into the cost (5 machine hours × £2 per machine hour). In Production department 2, the incidence of overheads is due to direct labour as opposed to machinery, thus the appropriate data is the six labour hours (not machine hours) and absorbed at the rate of £1 per labour hour, giving an absorbed cost of £6.

If the product output had required the use of only one of the productive centres, only one set of overheads would be required in costing the product. Nevertheless, this puts a great deal of emphasis on the need for planning as accurately as possible both budgeted activity levels and budgeted overhead amounts for functions and departments.

For estimation and pricing purposes, there would be similar principles applied for any administration and selling costs together with a percentage mark-up on all costs. However, for stock valuation purposes in the balance sheet, the production costs outlined above would be used to value stock under absorption rules.

Head office charging for contracts

In large engineering organizations where a great deal of activity is undertaken off-site for the client, such as building and construction, mining and exploration, specialist services are employed from the head office to assist the on-site operations e.g. geologists and seismologists may be required. Because these skills are often manual, they are charged out to the contract on the time spent. The head office or department would therefore, determine a company-wide overhead charge or a department rate.

For example, head office costs at Lang Engineering Ltd are £2,000,000 including all building and office costs, engineering and management's salaries, mainframe computing facilities and payroll. The overall charge is based on the budgeted level of activity for existing and future contracts of 4,000 labour-days. The rate charged to each contract would be £2,000,000/4,000 i.e. £500 per labour-day of head office involvement.

Batch costing

The principles of batch costing are, in most respects, identical to job costing. The major difference between them is that job is expressed in terms of the singular, and batch in terms of plural. Batch costing involves the replication of a discrete quantity of units in a single production run, such as batches of promotional material in the printing industry.

Contract costing

Although job costing is common in many manufacturing and service industries, many other engineering activities are more commonly costed on the basis of individual contracts. Civil engineering and shipbuilding industries commonly contract engineering services, and indeed, engineering activity within an organization may be contracted to other internal users, where business disciplines such as pricing and cost control will reflect the financial efficiency of the activity.

Generally, there are many similarities with the job costing referred to earlier. The major difference lies with the relatively long duration of the activity or project which often spans more than one accounting period, for example, construction, aircraft and marine engineering projects.

There are other characteristics that distinguish contract costing from job costing:

1 There is a higher proportion of direct costs, due to the fact that resources used are often specific, or easily traceable, to the contract; particularly if

the contract is undertaken on-site, e.g. site vehicles, contract engineers' salaries, power usage.

2 The corollary of the above would be a commensurate reduction in indirect costs. The need for the absorption procedure referred to earlier is substantially reduced.

3 Greater cost control often has to be exercised through supervisory management. Investment in materials and WIP, plant and machinery, health and safety and security procedures can be substantial.

4 With both jobs and small contracts the measurement of profit will occur when most of the revenue is realized in the same year. Where large contracts are involved, there may be a number of accounting periods before the sale is realized. However, independent assessors can estimate the value of the work undertaken to date and a percentage of the agreed valuation would be paid in cash and profit, to date, measured.

Firms embarking on a product offering likely to take more than one financial year from origination to delivery will undoubtedly employ some form of contract costing system. The method could be either absorption or marginal costing, though the former would be common if they were to conform to the rules of external reporting.

The overall account for the contract with the customer could be regarded as a profit and loss account. The major difference is that because the revenue will not be received in the current accounting year, a procedure has to be followed allowing for an estimate on the profitability of work to date. The problem, unlike conventional repetitive production and engineering, is that effectively WIP lies around in the balance sheet for the period of time of the contract. As there is technically no revenue to match the costs until the end of the contract, no profits could be measured and the capital providers will be leaving themselves exposed to large risks if reliant on some profit appropriation through dividends etc.

Worked example of contract costing

Grinders Mill won a major contract for a civil engineering project. The contract value was £4 million. To date, at the end of the current accounting period, the costs incurred were £2.2 million. It is estimated that another £1 million will be required to complete the project. Independent certification reflecting value of the work done to date was £1.65 million. Ten per cent of the contract price will need to be set aside for rectification and quality guarantees after the scheduled contract completion.

In measuring contract profit the following stages have to be undertaken per period until the completion of the contract. If profits are not taken periodically, there will be a reluctance to invest by the capital providers.

1 Calculate the estimated contract profit at this point:

	£m	£m
Total contract value		4
less		
Costs to date	2.2	
Estimated costs to complete	1.0	
Rectification	0.4	

$$= \text{Total estimated contract costs} \qquad \frac{3.6}{}$$
$$\text{Estimated profit (or loss)} \qquad \frac{0.4}{}$$

2 Determine the overall profit to date using the following formula:

$$\frac{\text{Certification of work completed}}{\text{Total estimated contract cost}} \times \begin{array}{c} \text{estimated contract} \\ \text{profit} \end{array}$$

$$\frac{\pounds1.65 \text{ million}}{\pounds3.6 \text{ million}} \times \pounds0.4 \text{ million} = \pounds183,333$$

3 The profit to date of £183,333 is cumulative over the life of the project. If the profit taken up to last accounting period was £120,000 then the profit for the current period is:

	£
Cumulative profit to date	183,333
Profit taken up to last period	120,000
Profit for this period	63,333

4 Suppose the value of the work deteriorated and/or the costs increased beyond those planned. If the cumulative profit in the next period fell to £160,000, a loss would have to be reported in the next year's PLA of £23,333.

Marginal costing

The use of marginal costing is primarily for 'short-term' decision-making. It is not acceptable for external reporting formats – only to assist the internal decision-making of engineers and managers.

The business organization's own resource constraints may only permit one costing approach. Because of the need to comply with accounting standards (SSAPs and FRSs) and the legal requirement to report externally, the costing system may follow absorption principles for internal management consumption. However, size of organization, complexity of markets and product range often justify investment in marginal costing systems and information technology to assist decision-makers rather than simply 'keep score'.

Marginal costing terminology

Fixed costs

These are costs that do not vary with decisions to increase or decrease output or activity in the short-term. These would include the Council rates payable on premises, salaries paid to existing managers etc.

Variable costs

Costs that vary in proportion to changes in output or activity over the short-term. These would include materials used up in manufacturing, direct labour etc.

Contribution

The difference between sales revenue and the variable costs of a product or service contributes towards paying the fixed costs of the business.

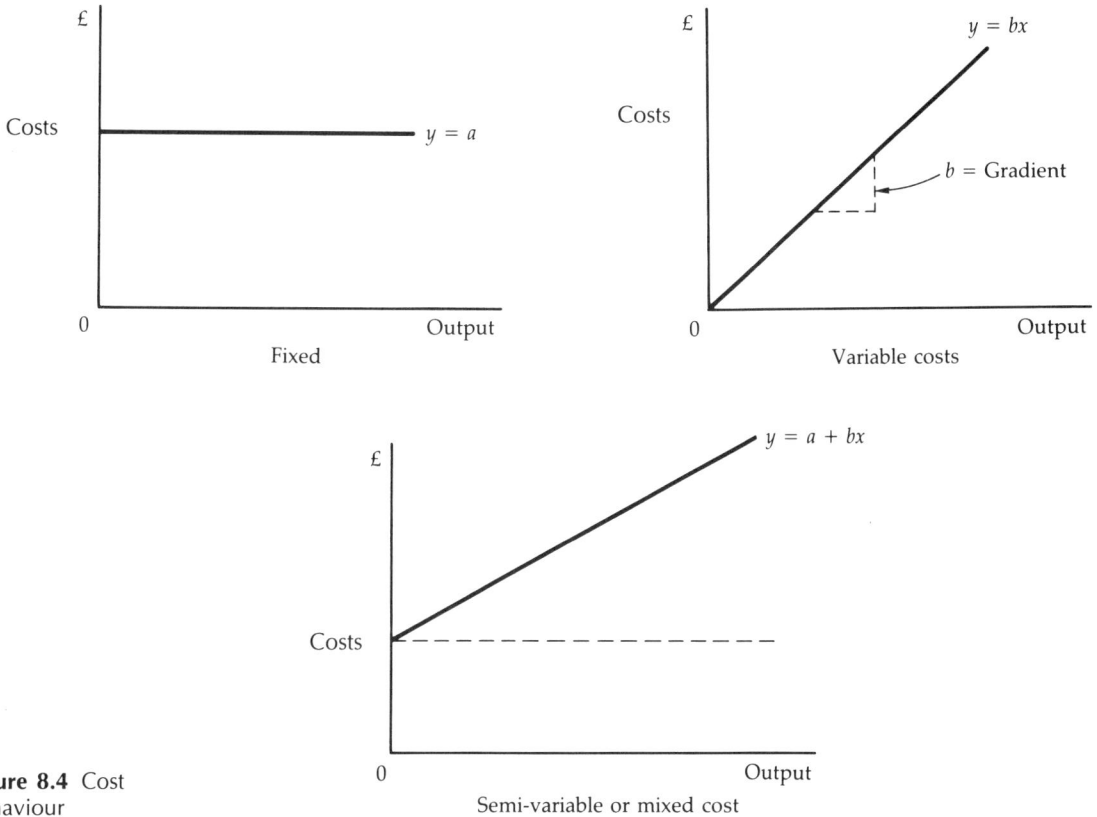

Figure 8.4 Cost behaviour

These terms are not used in conventional financial accounting for external reporting purposes (though many accounting texts do use the terms within the context of absorption costing to test the problem-solving abilities of accounting students).

Both fixed and variable costs can be expressed graphically as above. The vertical axis shows cost and the horizontal axis represents the level of activity or output and the key variables are represented by the following notation:

y = Total costs
a = Fixed costs
b = Variable costs
x = Output/activity

The semi-variable cost is a combination of both fixed and variable costs. The gradient, b, is a straight line representing the variable cost per unit of activity; the fixed cost is represented by a; x is the independent variable, and y is the dependent variable, cost. In the diagram above both expressions are combined, $y = a + bx$ (or alternatively expressed as $mx + c$).

The total cost expression of a product, department, function and organization can generally be summarized in this equation.

If costs have a high correlation with a level of activity then these will be considered variable. If costs have little or no correlation with activity in the short-term then they are considered as fixed costs. This does not mean that firms cannot avoid the cost in the longer-run i.e. fixed in perpetuity. It simply means costs are not avoidable in the short-term.

Breaking-even and profit planning

Where an organization is providing a standard product, marginal costing can prove a useful short-term technique for evaluating the relationship between certain types of costs, profits and volumes.

Broadly, all costs of an organization (or department, sub-unit, etc.) can be expressed as Total Costs (TC). Under marginal costing these total costs must be the sum of both fixed costs and variable costs. Because fixed costs do not vary with output in the short-term then the absolute total is usually the appropriate expression. From a management perspective this would imply that they are costs which cannot be avoided in the short-term by their decisions alone. However, variable costs are incurred because there is some measurable output and/or activity. The management decision, at the margin, is to increase or decrease that activity. Thus the variable costs are expressed as a function of the unit of output.

TC = Total Fixed Costs (FC) + Total Variable Costs (VC)

where variable costs are a function of the unit variable cost and price per unit, then

TC = FC + (VC/unit × Output)

where output is the independent variable under consideration for change at the margin by management.

This relationship can be expressed graphically as in Figure 8.5.

The vertical axis is costs as measured by £. The horizontal axis reflects output: units, litres, boxes etc.

The equation is a straight line: $y = a + bx$
where y = dependent variable (i.e. total cost)
  a = the fixed costs (constant over short term)
  b = the gradient (the variable cost per unit)
  x = independent variable expressed as a level of activity (i.e. number of units produced or level of service provided)

In addition, if a revenue line is introduced, total revenue (TR) = price per unit × output = Px
where P = price of the output to the customer
  x = the independent variable of activity (i.e. number of units sold).

Contribution

When variable costs are subtracted from the sales revenue at a given level of activity, the difference is known as contribution. Profit is found by deducting the fixed costs from the contribution. The term contribution is specific to marginal costing, in the same way that gross profit is to absorption costing.

The intersection of total revenue and total costs is where TR = TC, or, using the algebraic notation:

$Px = a + bx$

Figure 8.5
Conventional break-
even

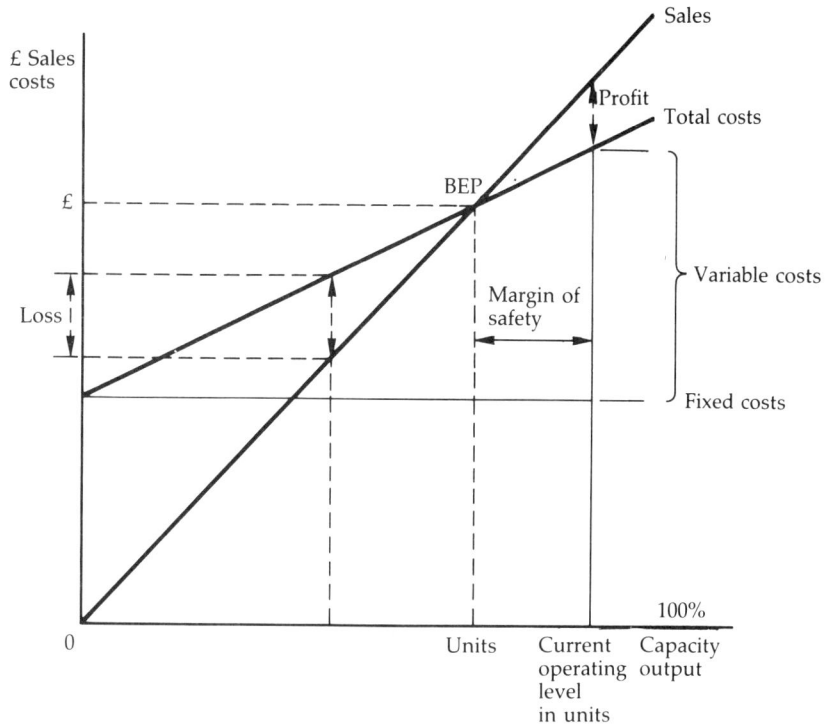

rearranged to give

$$x(P - b) = a$$

$$x = \frac{a}{P - b} = \frac{\text{Fixed costs}}{\text{Contribution/unit}}$$

This is known as the break-even point (BEP) and indicates the level of output at which sales revenue = total cost. A further refinement of the break-even point allows profit targets to be planned:

TR > TC = profit
let M = profit

Then

$$\text{TR} - \text{TC} = M$$
$$Px - (a + bx) = M$$

rearranged,

$$x = \frac{a + M}{P - b} = \frac{\text{Fixed cost + planned profit}}{\text{Contribution/unit}}$$

The above can be rearranged to solve for any one of the variables as long as the others are known. For example if the break-even formula of fixed costs/contribution per unit is multiplied through by price, then:

189

$$P \times x = P \times \frac{a}{P - b} = \frac{a}{(P - b) / P}$$

Where $(P - b)/P$ = Contribution per unit/Price per unit, this can be expressed as a percentage ratio known as the contribution/sales ratio:

$$\text{Break-even} = \frac{\text{Fixed costs}}{\text{Contribution/sales percentage}}$$

This means that the break-even point can be expressed in terms of revenue i.e. the intersection reading taken from the vertical axis of the graph.

The technique will have less relevance to an engineering organization that provides a low volume customized/bespoke product(s)/service(s) designed to technical specifications for its customers such as boiler fabrication. However, a software house generating large volume commercial software products to customers with non-proprietary hardware might use this technique to estimate the effect on contribution of possible combinations of sales volume and price which would be achievable in the marketplace.

Scientific measurement of the costs may indicate that there is a non-linear relationship between output and unit cost. Superior mathematical models could be handled with computers, but it must be recognized that the expression is an aid to decision-making and needs to be taken in context with other criteria. Because many decisions are made at the margin in the short-term, the relevant range could act as a guillotine on a marginal costing graph. This would imply than any decision to move beyond either extreme of the graph is prone to greater error margins and unreliability.

Further, for marginal costing to have any significance in aiding decision-making, there needs to be a systematic examination of the cost behaviour of each type of expenditure. This will mean a re-classification of the conventional costing terminology.

Limitations to marginal costing

The use of technique must be conditioned by an understanding of its limitations:

1 The assumptions in the analysis assume linearity off sales, fixed costs and variable costs.
2 The relationships are short-term and probably inappropriate over longer time horizons.
3 There is an assumption that the activity chosen is the sole determinant of costs and revenues changes.
4 The relationships hold within a reasonable range of the average, beyond which there is less reliability i.e. the relevant range referred to earlier.

Comparison of absorption and marginal costing methods

Absorption	*Marginal*
Direct materials	Variable materials
+ Direct labour	+ Variable labour
+ <u>Production overheads</u>	+ Variable production overheads
= Absorbed cost of production	+ <u>Variable non-production overheads</u>
+ Non-production overheads:	= Variable costs
	+ Fixed production overheads
	+ Fixed non-production overheads
<u>Total cost</u>	<u>Total cost</u>

1 The absorption costing format concentrates on the transformation of the raw material through to the finished good to be dispatched for sale. The accumulation of costs is built up through production. Once the cost of production has been determined then, after adjusting for stocks, the cost of sales can be determined. Other functional costs will be added in order to arrive at total costs for the firm.

2 The classification of direct costs includes both variable and fixed costs. For the most part the direct costs tend to be variable. Raw material costs are direct under absorption costing and variable under marginal costing. Direct labour is often treated, traditionally, as variable. The rule is to treat as variable only those costs that vary in proportion to the level of activity.

3 Regardless of the functional areas of the business, variable costs are all those costs that will change with output or activity changes. Sales commission on the sale of the product would vary with volume of sales making it a variable cost. Direct material costs are incurred because of production activity and are, similarly, variable.

In the Profit and Loss statement there are also differences:

Absorption	*Marginal*
Sales *less* absorbed product costs = Gross profit	Sales *less* all variable costs = Contribution
less non-production overheads	*less* all the fixed costs
Net profit	Net profit

Marginal costing applications

Relevant costs

Rational decision-making invariably involves comparing identifiable options expressed financially. When these options are compared, some of the costs may be common to all options and thus considered as irrelevant. Other costs may have been incurred already, known as sunk costs, and are also irrelevant. Only those costs that will be incurred if the option is undertaken should be considered. Relevant costs include both fixed and variable costs.

Dropping product line

If costing data is presented in an absorbed format i.e. where costs have been arbitrarily allocated, some products or projects may appear to be unprofitable:

	Product A £000's	Product B £000's	Total company £000's
Sales	200	150	350
Variable costs	120	130	250
= Contribution	80	20	100
– Fixed costs	60	30	90
Net profit	20	(10)	10

There are two cases here:

1. If management decided to drop Product B and all the fixed costs for the firm are unavoidable, then the £30,000 fixed costs allocated to Product B will simply be transferred to Product A and the company, in the absence of any replacement of Product B, would make a loss of £10,000. In other words, the contribution of Product B would be lost with no commensurate savings in fixed costs to offset this.
2. Some of the fixed costs may be directly attributable to the product but could be avoided if Product B has to be dropped, if we assume the breakdown of fixed costs for Product B were as follows:

	Product B total	Avoidable by dropping Product B
Administration	9	6
Marketing and advertising	16	13
Rent, business rates	3	1
Depreciation	2	2
Total:	30	22

This would suggest to engineers and managers that Product B should be dropped because the costs relevant to the decision: £22,000 fixed costs saved in this case are greater than the contribution of £20,000 lost. £8,000 of fixed costs are irrelevant to the decision.

£
22,000 fixed cost saving
20,000 contribution loss

2,000 net profit gain

Make or buy decisions

Many organizations in both the private and public sector are facing these decisions frequently. For example, in the manufacturing environment, in-house components manufactured would need to be compared with the bought-in purchase price. Generally, if the latter is more expensive than in-house manufacture, the company should continue to manufacture. Equally, in the public sector, local authority service are having to compete with outside private sector service organizations (where assuming homogeneity of non-quantitative factors such as quality) and costs may play a large part in securing the contract.

Again, there are two cases:

1. If the cost of the component is presented in absorbed format then it is the relevant costs that need to be compared for decision-making purposes.

For example, Speck-Eng Ltd produces a small component for its machine tools. The total cost of this part, on the basis of normal production of 5000 units, is as follows:

	£/Unit	£/Unit	£/Unit
Variable costs:		15	
Fixed costs:			
Depreciation	4		
Administration	5		
Salary	2		
Total	11	41	26

If an outside supplier offered a similar quality component (with similar service support) for £19 a unit, Spec-Eng Ltd will need to consider whether to continue making or buying in from outside. On the face of it, buying from outside would be cheaper.

If none of the fixed costs were avoidable by ceasing in-house manufacture, then the relevant costs will be the variable costs saved compared with the buying-in price – £15 compared with £19. The rational decision would be to continue manufacturing, as an additional £4 per unit would have to be spent if the component were bought in.

Alternatively, if £6 fixed costs were also avoidable then it would be cheaper to buy-in from outside – even if there was no profitable alternative to replace the existing manufacturing capacity.

i.e. £15 + £6 > £19

Therefore: There would be a saving of £2 per unit by buying in.

2 Capacity

If not making the component meant that the facilities could be released for a profitable alternative, consideration must be given to the profit arising from the replacement in deciding whether to make or buy. For example, if buying-in cost £19 and relevant costs of manufacturing the component were £15, but by using those same facilities contribution could have been earned of £5 per unit on another product, there is an opportunity cost associated with the use of the facilities which makes the buying-in option attractive.

£15 > £19 – £5 i.e. Buy-in.

Opportunity costs

If B.O. Sullivan, an electronics engineer, considered giving up his job which currently earns him £22,000 per annum and investing his savings of £100,000 (currently invested at 10 per cent per annum) in his own sub-contracting firm, which will, according to the business plan, generate £35,000 per annum after tax, advise him on the financial consequences of his action.

	£	
By starting his own firm, net profit after tax would be	35,000	
However, the opportunity cost of salary foregone	(22,000)	
and, the opportunity cost of interest foregone	(10,000)	
Net gain per annum		3,000

Hence, the financial consequences of undertaking another option need to be offset against the benefits foregone from the existing position.

Activity based costing

The costing system employed by a firm can take on greater significance with companies that are following particular strategies. If the competitive strategy is to produce a generic product with little or no differentiation from other products, costing is likely to provide a very useful tool. If the business wanted to distinguish the product in the marketplace then the price sought from customers would need to cover the additional costs needed.

Activity Based Costing (ABC) differs from the conventional method of absorption costing. The absorption technique identified earlier organizes its collection of costs through production and service cost centres. Any indirect costs of engineering and other service cost centres were allocated, apportioned and absorbed into production output. The problem with the absorption approach is that although it helps cost a product it doesn't lend itself to assisting the management process of controlling and reducing costs. It was stated earlier, for example, that the incidence of overheads would be sufficient to be absorbed over the base that caused the overall incidence. For example, canteen costs might be apportioned on the basis of the number of employees in each production and service cost centre. That many of the employees may have alternative arrangements to the extent that some cost centres may not use the facilities at all may not affect the overall absorption process.

Absorption costing measures the use of the resources through an expression of cost per/unit of output. However, it has severe limitations for planning and control purposes. ABC attempts to overcome some of these problems by attaching costs to activities. It enables the organization to identify what activities drive what costs.

ABC attempts to determine product costs with less arbitrariness than absorption. High volume standard products are likely to incur the bulk of the overhead costs and of absorption, while the short-run bespoke product will attract less. However, in reality, the smaller volume product may require longer and more expensive set-ups, tooling, engineering and production support.

ABC relates the support overheads to products not through volume but by specific cost factors known as 'cost drivers'. Cost drivers are support activities which use resources in order to produce the final output.

Cost drivers	Example of costs driven
Number of procurement orders	Buying department, stock-holding
Number of engineering operations activities	Technical and production, planning, stock-holding
Number of production runs	Quality and inspection, production planning, scheduling, set-up, tooling
Number of despatch orders	Distribution departments, customer services, after-sales

Figure 8.6 Traditional absorption costing and ABC contrasted

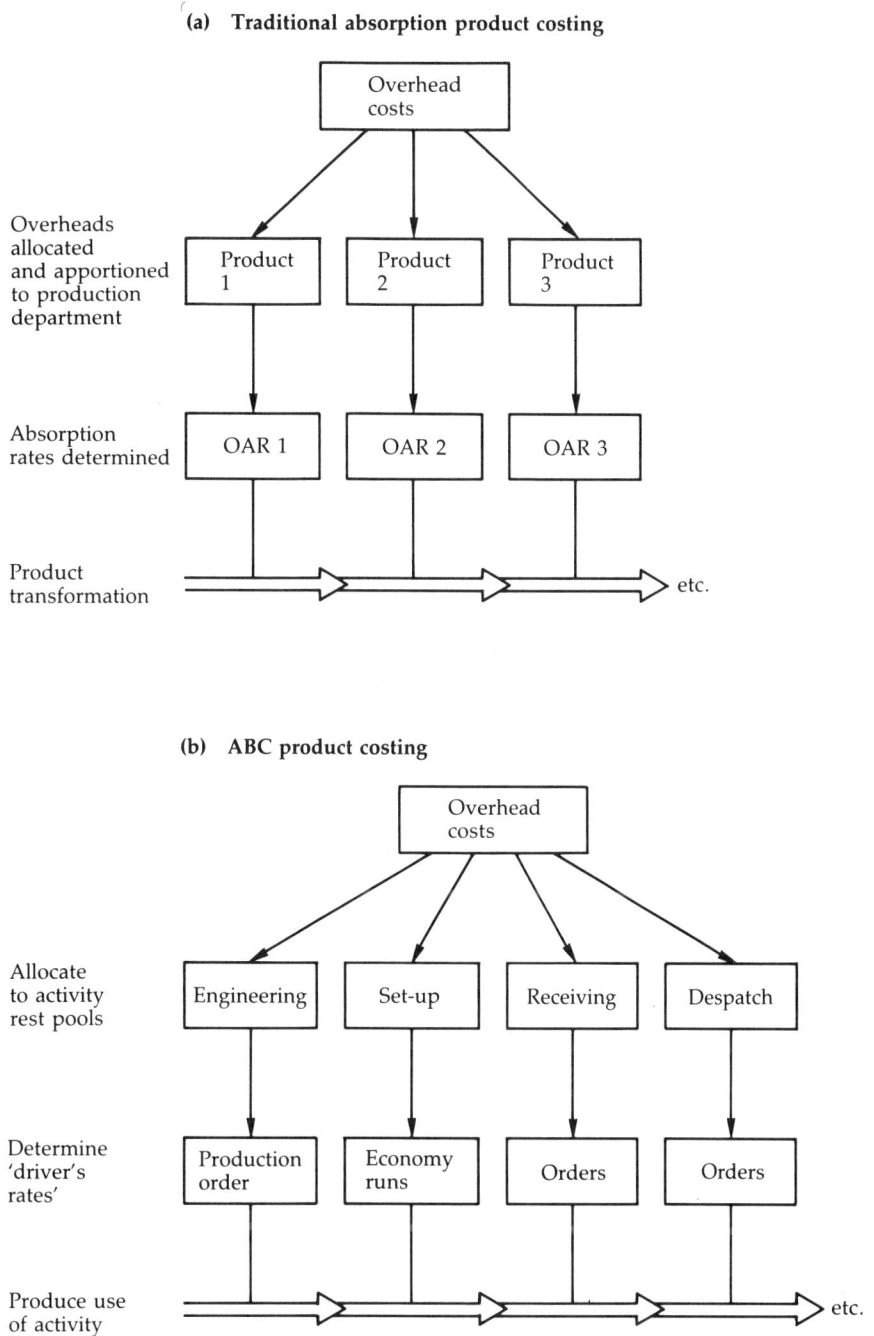

(a) Traditional absorption product costing

Overheads allocated and apportioned to production department

Absorption rates determined

Product transformation

(b) ABC product costing

Allocate to activity rest pools

Determine 'driver's rates'

Produce use of activity

Essentially, the operations of a typical manufacturing environment are regarded as a continuum from design – procurement – set-up – production processes – distribution – sale – customer support.

The firm is regarded as an integrated set of activities which attract resources. There are two groups of activities: value added (VA) and non-value added, (note that accountants use this term differently to marketers; see Chapter 2). Each activity might attract resources from more than one functional area. For example, set-up costs may require materials handling, tooling and indirect costs. A costing rate driver is then found for each individual activity expressed as a measure that is constant over the activity. If set-ups are standard, then the costs can be absorbed over the number of set-ups. If they are not standard then set-up might be better expressed in terms taken per set up time.

Value added (VA) activities add value to the product through labour, materials and machines. Non-value added (NVA) activities do not add value and could be removed from the cost: such as set-ups, rework and material handling. NVA may not be avoidable – but they could be reduced over time. Modern production systems attempt just that through CAM. This would leave management attention to focus on VA costs for control.

Because output may vary in its consumption of resource to each activity undertaken, by following the process flow, costs can be determined at a unit, batch, product line, and by market segment or customer. To date, the evidence supporting ABC is not overwhelming. Some regard it as an improved methodology for determining the product cost than conventional absorption. However, this may not always lead to improved decision making.

Value added and competitiveness

A firm using the value added approach is in a much better position to maintain a competitive strategy. By using the marketing techniques referred to in Chapter 2, and the costing techniques referred to in this chapter it is possible to estimate:

1 The value of any production or marketing activity to the consumer;
2 The cost of providing that activity.

An organization wins a competitive advantage if it can provide the former in a way different to another company (providing that the cost is not greater than the consumer, or segment, is willing to pay) or by reducing the latter to achieve cost leadership.

Increasingly, firms analyse their competitive strategies in terms of the value chain technique first developed by M. Porter (see references in the further reading section of Chapter 3). The technique itself is rather beyond the scope of the book, but principally involves examining costs and value added in five primary areas:

- Inbound logistics
- Operations
- Outbound logistics
- Marketing and sales
- Service

and comparing the value added to the margin obtained. Clearly, the value added must be greater than the margin for the company to survive and develop a competitive advantage, as well as pay interest, wages and salaries etc. Competitive advantage in any of the five areas can be further developed by the application of support activities, defined as being:

- Human resource management
- Technology
- Procurement
- Organizational infrastructure

It is argued that an organization possesses cost leadership if the overall costs of the value chain are less than its competitors. If this advantage can be sustained, then the organization will have a strategic advantage in the marketplace.

Chapter summary

In this chapter the two principal costing techniques have been explained and demonstrated. Examples suggest that absorption costing is important to ensure that the costs of a firm are at least met by sales receipts, and to determine where costs are incurred. Marginal costing is more appropriate when applied to rational decision-making circumstances, to ascertain whether or not additional activity would contribute towards the fixed costs of the firm in the short-term and is slowly being adopted. As the use of the latest technology spreads in the UK, ABC is gradually becoming acceptable to engineering and production firms. Although the measure of product costing will be greatly improved, its performance in improving management decision-making has not been entirely convincing to date.

Further reading

Arnold J. and Hope T. (1990), *Accounting For Management Decisions*, provides a clear and cogent development of marginal costing methodology in short-term management decision-making. Ryan B., *Management Accounting: A Contemporary Approach* (Pitman, 1985) makes the subject matter particularly readable and intelligible. Blommaert, A., Blommaert, J., and Hayes, R.S., *Financial Decision-Making* (Prentice Hall, 1991) takes a wider European perspective to management decision-making. An integrated case study approach is taken in Wilson, R.J., *Financial Analysis* (Cassell, 1987).

References to value chain and competitive advantage can be found in the further reading guide to Chapter 3.

Tutorial exercises

8.1 Consider the discussion between Alex and Bill on page 169. Who was correct in this instance?
8.2 (a) Given a selling price of £20 per unit, fixed costs of £6,000, and variable costs of £16, find the break-even point in units.
How many extra need to be sold to make a profit of £2,000?
(b) If new technology became available, which reduced variable costs by £2 per unit, but raised fixed costs to £2,000, what would be the new break-even point? At what output (sales) would the new technology improve the company's financial position?
8.3 Grinder's Mill have produced the following data for their car parts range of products.

	19X2 £	19X3 £
Sales	400,000	500,000
Operating costs	380,000	440,000
Net profit	20,000	60,000

With no inflation, the selling price for both years remains at £20 per unit. All variable costs are regarded as a linear function of output. The output in both years was at full capacity of 30,000 units.

Required:

(a) Calculate the volume of sales in £s needed to break even.

(b) Determine the unit cost when the firm is operating at 80 per cent capacity.

(c) What volume of output is required to make a profit of £60,000.

(d) 19X4 is anticipated as being a better business year and management are considering a number of possible situations. Determine the profits for each case:

(i) Unit sales increase by 10 per cent.

(ii) Unit prices increase by 10 per cent but volume sales fall by 5 per cent.

(iii) Unit prices fall by 10 per cent and volume sales increase by 20 per cent.

(iv) Improved engineering quality will increase variable costs by 10 per cent.

(v) Increased marketing activity through advertising increases costs by £50,000. Unit prices are increased by 20 per cent. Volumes increase by 5 per cent.

8.4 OTT Transport specialize in long distance haulage of engineering equipment. OTT's revenue is determined by UK mileage. Total budgeted mileage for next year is 20,000 revenue miles and summarized data is as follows:

	Per mile
Average selling price	8
Average variable costs	5
Fixed costs	£600,000

Required:

(a) Determine budgeted net profit for the budgeted year.

(b) Determine the effect on profit of the following independent changes:

(i) 20 per cent increase in revenue miles.

(ii) 10 per cent increase in sales price.

(iii) 20 per cent increase in variable costs due to petrol increases.

(iv) 10 per cent increase to fixed costs for salaries and overheads.

(v) 5 per cent increase in revenue miles and 25 per cent increase in fixed costs due to advertising.

8.5 B. Keoghy is rather concerned that the cost per unit given by his accountant for British Standard Extruded Tubing is not reflecting the need to fill unused capacity with special orders to new customers overseas. Recently, a contract order worth £72,000 for the same specification from Trinidad for 3000 units was rejected by his engineering sales because it was unprofitable. Maximum output is 12,000 units per month. The output levels in the past three months were:

Month	Output levels (units)	Total cost £	Cost/unit £
April	10000	250,000	25.00
May	7000	205,000	29.30
June	5000	175,000	35.00

Required:

(a) Plot the above data and determine the fixed and variable costs.

(b) Evaluate whether or not the contract should have been accepted on the information given. What additional information might be required?

8.6 (a) Rila Engineering Ltd is considering the launch of a newly developed precision tool in the next financial year. In manufacturing the product there is a choice of three alternative production processes X, Y and Z that need to be considered by the engineering management. The anticipated selling price will be £100 per unit:

	X	Y	Z
Variable costs per product (£)	75	80	85
Fixed costs (£)	150,000	100,000	60,000

(The capacity constraint on each process is one million tools)

Required:

(i) Evaluate the three mutually exclusive alternatives, both numerically and graphically, using cost-volume-profit relationships.

(ii) Identify any other significant factors that would be taken into account before the decision is made.

(b) J. MacAdam has recently left a firm of non-destructive-testing engineers to set up his own business. His anticipated monthly costs are:

	£
Secretary's salary	1000
Rent of industrial unit	600
Heating and lighting	250
Telephone	150
Depreciation	100
Total costs	2100

He anticipates earning £2,400 per month and fully expects having to invoice clients for 150 hours of his time each month.

(i) Calculate the rate per hour Mr MacAdam will need to charge clients.

(ii) Determine how much he would earn if he worked only 100 hours in a given month. State any assumptions made.

(iii) Consider the limitations of this approach if an ex-colleague decided to compete on price?

8.7 Turbo-X engineered a successful prototype carbon bicycle and manufactured 100 as a limited edition in the monocoque style in September 19X9. The costs incurred were expressed in absorption form as follows:

	£
Direct materials	10,000
Direct labour	20,000
Production heating	5,000
Production energy	4,000
Production indirect labour	2,000
Rent (1/2 production)	1,000
Business rates (1/3 production)	1,500
Depreciation (straight-line, 3/5 production)	500
Interest costs	400
Selling and distribution costs	2,600
Total	47,000

Required:
(a) Calculate the production cost per bicycle.
(b) Calculate the total cost per bicycle.
(c) Determine the total fixed costs and variable costs per unit.

8.8 Fast-chip produces a standard IBM specification printed circuit-board. One direct labour hour is required for each unit at a rate of £10 per hour. The direct material cost is £12 per unit. The selling price is £30 each.

For the month of January the actuals were:

Sales	1,800 units
Production	2,000 units
Direct material costs	£24,000
Direct labour costs	£20,000
Fixed manufacturing overheads	£6,300
Fixed administration overheads	£2,200
Variable selling and distribution	£1,800

The budgeted data for the 12-month year including January was:

Production	18,000 units
Sales	16,000 units
Fixed manufacturing overheads	£72,000
Fixed administration overheads	£26,400
Variable selling and distribution	£16,000

Required:
(a) Calculate the profit for January using absorption costing.
(b) Calculate the profit for January using marginal costing.
(c) Determine the budgeted break-even point for the year.

9 Budgeting

The chapter can be considered, at the time of writing, in the context of the general UK current economic climate of falling outputs, falling investment levels and weak confidence in the future. Although many commentators prophesy recovery and growth, it is still clear that the UK will take some time to recover from the recession of the 1980s.

A decision to postpone investment decisions and possibly make employees redundant is often taken following a strategic review (see Chapter 3) and a subsequent budgeting exercise. Many engineering businesses, though profitable, are facing severe liquidity (cash) problems that can cause imminent failure. The paradox is often expressed '...but the accountants told us we were profitable' shortly after administrators of the firm have been appointed by a court because creditors can't be paid.

Although the final financial statements of profit and loss accounts and balance sheets described in Chapter 6 didn't specify cash statements, this should not be taken to imply that they are not important to the engineering and business manager. Cash flow statements for engineers and management are essential. Further, the frequency with which they are generated internally is essential for ongoing planning and control.

Figure 9.1 is an extract of the original model in Chapter 6 and continues the management accounting aspects from t_0 to t_1 with the emphasis on planning and control. This chapter will describe the type of budgeting process within a business organization and some of the variations that might be found in others. Unlike financial reporting to external parties there is little or no legislative framework to conform to – it is entirely optional whether a firm budgets or not. Although the complexity of

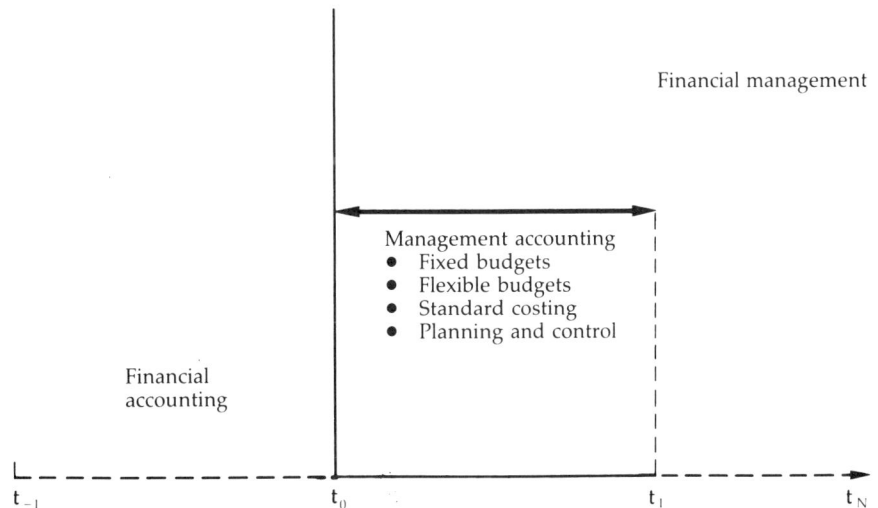

Figure 9.1 Context of management accounting

Financial management

Management accounting
- Fixed budgets
- Flexible budgets
- Standard costing
- Planning and control

Financial accounting

t_{-1} t_0 t_1 t_N

modern organizations and their interdependence would suggest that firms would find it very difficult to survive without some form of budgeting.

Budgeting involves the preparation of quantitative and financial plans based upon the overall organizational objectives over a future period. This is usually taken over one year and broken down into either twelve monthly or thirteen four-weekly cycles. This is clearly a major operation in any organization involving engineers and other managers. The end product of the budgeting process is the blue-print for the organization's business activities over the next financial year measured in financial terms. The budget for the organization should encapsulate the resource acquisition of the firm (sales of products and services, loans etc.) and its intended resource allocation (departmental overheads, product costs and investment) over the short-term – usually up to one year ahead.

The two key outcomes for senior management will be the budgeted final accounts (budgeted profit and loss account and balance sheet) and the budgeted cash-flow statement. Although cash-flow was briefly alluded to in Chapter 6, its importance to the continued existence of a business organization cannot be overstated. It is quite possible through the accounting accruals-based measurement system to provide a profit while simultaneously experiencing cash deficits. The inherent dangers of not being able to meet short-term debt obligations can lead to liquidation. An example later on in the chapter demonstrates just how easily it can happen.

In large companies the ownership of the budgeting process may be shared throughout the organization; though not necessarily evenly. For example, in a hierarchical organization, senior managers have to make projections that they feel would satisfy their shareholders while simultaneously recognizing the current and projected trading climate within which they operate. Middle managers would then be tasked with controlling the expenditure of resources with agreed budgets on a routine basis. Any significant deviation between projected and actual expenditure needs to be highlighted and the appropriate management action taken. However, in some organizations, middle management may have much more flexibility in the way budgets are set and spent.

Budgeting is related to past financial events, current activities and long-term developments. This would suggest it has a link with the recent historic records of business activity i.e. financial accounting. Simultaneously, it should be sufficiently flexible to respond to current events such as changing interest rates and responding to competitive actions. Furthermore, it needs to reflect the overall short-term strategy of the company. This must be consistent with the longer term marketing and sales plans of the company, and when this is not the case it will be impossible to co-ordinate the firm's activities. The actual task of putting financial details and expenses to the long-term plans is normally undertaken by a budget committee.

Budget timing

Normally budgeting is undertaken on an annual basis. Figure 9.1 implies that budgeting activity can be separate from the production of the financial accounting statements. There is considerable overlap and the two systems run in parallel with actual financial data being used for both control purposes and for extrapolation into budget reviews. If the first control period of the budget year is significantly different from that budgeted because of extraneous factors such as an interest rate rise affecting the engineering construction industry, then it might prompt an overall management review of some of the assumptions in the original budget.

Budget committees

The constitution of a budget committee will vary from firm to firm. In small businesses there will be less need for a grouping of senior managers because there will be less interdependence. i.e. the owner will be the engineer, the sales managers, the production director and senior administrator combined into one.

In larger organizations, the budget committee will be made up of functional representatives with sufficient authority to commit their part of the budgeting process. For example, in a capital goods manufacturing business there will be senior representatives from engineering, production, sales and marketing. The role of coordinating the budget is often undertaken by someone with an accounting background.

Role of budgets

Although budgets are the financial representation of corporate objectives and plans, they do have several other useful roles to play.

Communication

The operations of organizations can be very complex. It is very difficult to maintain centralized control in any organization that has to respond to the competitive, technical and legal turbulence of the environment. The budget process can ensure that overall financial objectives are communicated through budgets to the various departments, functions and divisions of an organization.

Coordination

The sales force may be geographically removed from the day-to-day activities of the production units. The budget will enable overall framework of production targets being met to enable the sales force to undertake their day-to-day activities.

Authorization

Budgets usually go hand in hand with managerial responsibilities. In its simplest context it means that an engineering manager entrusted with the use of the organization's resources, be they other employees, buildings, materials and overheads are responsible for the judicious use of those resources and responsibilities. In Chapter 8 reference was made to cost centres. These cost centres are often used as a basis for monitoring and controlling expenditures. For example, trainee engineers would need to obtain authorization for travel from their line manager if this was not part of their usual responsibilities.

Objectives

Budgets are often the foremost major communication of business intent over the forthcoming budget period. Resources are actually being committed and their use recorded in terms of materials, time, manpower, building overheads, overseas travel and other resources. From a management and engineering perspective, these resources are used in carrying out the objectives of the firm through projects, departments and divisions.

Management control

Budgets provide the major method of exercising engineering and management control. Any capital expenditure authorized for engineering projects will be entrusted to a project manager. The project manager will probably be allocated these funds in phases such as design, implementation and testing. Not all of the funds will be provided in one go. The project will need to account for progress periodically and will be measured by key milestones achieved and the financial resources used in getting there. Using the budget model in Figure 9.2, if t_0 to t_1 were a one-year period, then there is a budget for the purpose of management control that may have the same inherent weaknesses as historical financial accounts. There is no indication of when the problems arose and the possibility that they may have been compounded, thus making it difficult to find the likely root cause and effect action.

Control periods can be expressed in quarters, calendar months and four-weekly cycles. Decreasing computing costs through distributed systems and networks enables shorter control periods.

The budgeting process

In this section we look at the process by which individual budgets are actually set, and examine an example. The process of budgeting will be dealt with by identifying the constituent parts of a manufacturing company in terms of resources to be acquired, stored, expended and revenue earned.

The approach is prescriptive for the purpose of exposition. In reality, however, each organization will have its own approach to budgeting dependent on the management style, competitive environment, technology availability and organizational skills.

Figure 9.2 provides a broad model of the approach taken. There are four elements that need to be considered:

1 Individual constituent or subsidiary budgets. These usually reflect a cost centre, product, departmental, functional or divisional approach.
2 Linkages between the various subsidiary budgets. For example, in a production environment it is important to recognize the capacity constraints and the ability to control physical and financial stock levels before committing to sales targets.
3 Final statements reflecting the coordination of the subsidiary budgets in financial terms. These are often known as master budgets and express the projected overall results of the company using:
(a) The balance sheet and PLA as frameworks for expression of overall results and
(b) A cash budget; normally called a cash-flow statement over the same period of time.
4 Once the budget has been approved by the senior management at managing director or board level then it is adopted for the forthcoming financial year.

Figure 9.2 is a simplified budgeting model for a manufacturing organization. It can be regarded in a time frame because often there will be a critical period when the budget takes its overall final shape only to undergo further refinements before it is adopted by the organization. Again, the process and modelling lend themselves to the use of computerized

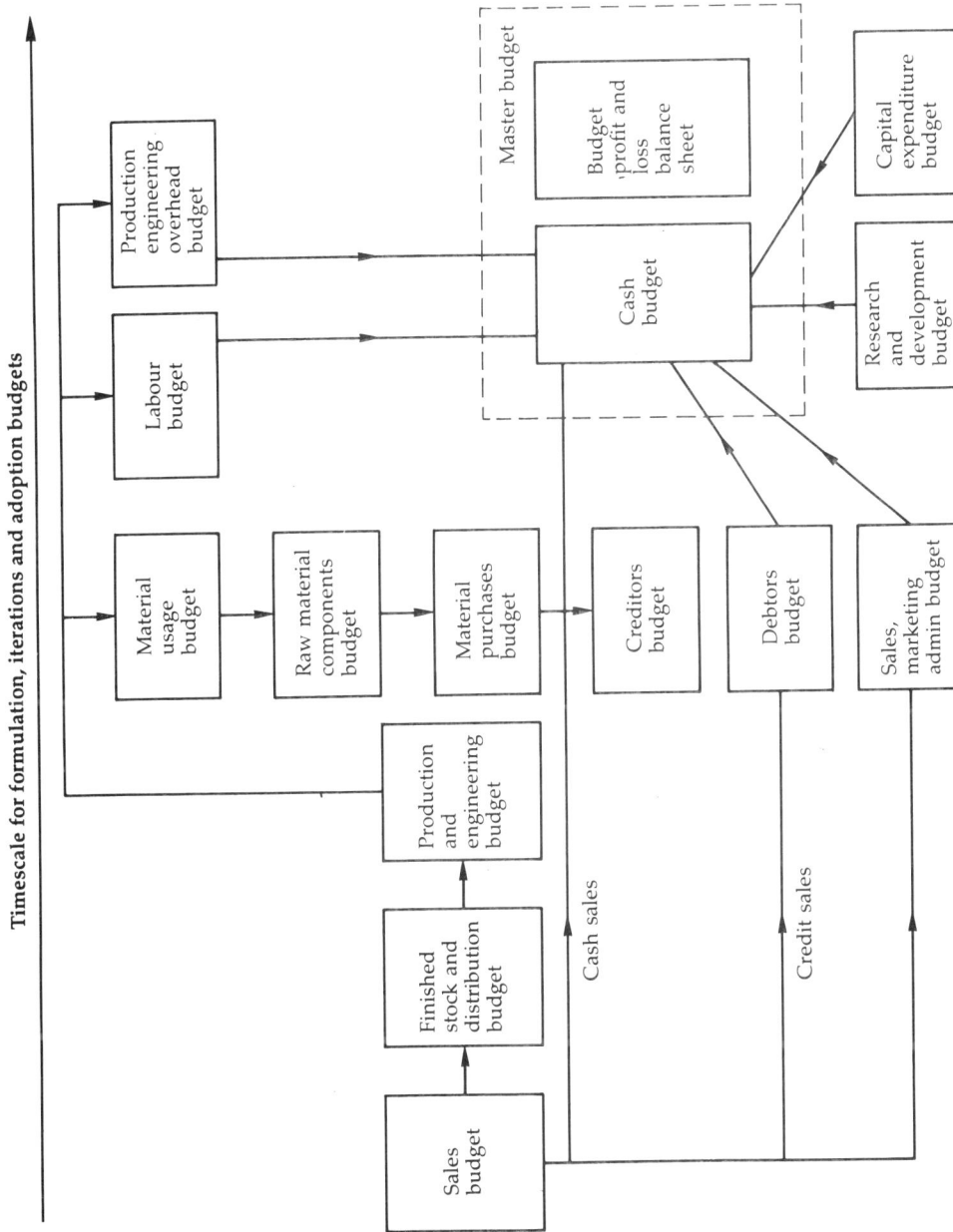

Figure 9.2 A budgeting model of an engineering and production business

information processing. Figure 9.2's constituent parts of the budgeting model will be examined.

Sales and marketing budget

The ownership of sales and marketing information rests with those active in customer and market interface. The salesforce's knowledge and opinion about future buying intentions, together with marketing intelligence, will indicate the range of product offerings, prices and sales volumes to be achieved. This information would need to be translated into a quantitative expression such as price and volume projections for each type of product and service provided. The anticipated level of credit sales will have a large impact on both cash flow and budgeted balance sheet.

Finished goods

The operational flow of goods through the distribution system is critical. If there are insufficient warehousing facilities available because of an increase in projected volume sales then distribution management would have to consider three possibilities:

1 To acquire additional distribution capacity through temporary sub-contracting or permanent capital expenditure.
2 To simply not meet the projected sales demand.
3 To change some element of the marketing mix to reduce the demand; i.e. raise prices or reduce the supply of products to less profitable channel chains.

Production budget

Once there is some understanding of projected sales and marketing intentions, and a recognition of the distribution issues, the production department can determine their overall plans and schedules over the budget period.

However, sales productions may not always drive the production plan. In a climate of growth, the production capacities may prove to be the constraint and the marketing and sales budget become a key dependency of the production budget.

In determining the overall production plan, there are a number of implications for managers of resources within the organization, which generate another series of subsidiary budgets:

1 Once production plans are known, there are implications on the acquisition of materials and components from suppliers. Buying or procurement managers would be involved in this exercise. A knowledge in advance of requirements will assist their negotiations with suppliers for prices, bulk discounts and quality.
2 There is some financial involvement because the quantities of materials and components to purchase will require financial management of stocks, creditors and debtors (commonly referred to as working capital management). Indeed, there will be stocks of raw materials that would need to be reviewed to determine the level of purchases required. For example, if the stock of a particular component were high, and the product which uses it is to be phased out, there will need to be a major review of the buying of this component. Some optimization technique, such as a sophisticated version of the economic order quantity, could be used to determine new order sizes and frequency.

3 Service department budgets such as engineering design, will also be reviewed on the basis of planned activity levels in the organization for the coming year. In some service departments, particularly in engineering research or development roles, there is no direct relationship with current production budgets.

4 Labour requirements can be assessed in advance of the budgeting year. If products can be determined by volume then the labour requirements in labour hours can be established. Costs can be determined by using the existing wage rates, and any proposed increases. In addition this exercise often identifies shortages and surpluses in skill requirements and establishes training and recruitment needs. In times of surplus capacity, voluntary or enforced redundancy, part-time and short-time working might be initiated.

Non-production overhead budgets

There are other non-production overhead budgets to consider such as sales, marketing and administration. The model links these budgets to sales activities, implying that the costs are driven by the proposed sales activities. Some marketers would argue that it is when the activity levels are falling that marketing becomes of greater significance (for example, advertising expenditure ought to rise in a recession). Engineering and non-production budgets indirectly related to the level of output would include:

1 The research and development budget: only relevant in an organization that has a commitment to these activities. Businesses using substantial engineering expertise such as oil, chemicals, food processing industries etc. are likely to identify separate budgets (in some cases the R&D function is given company status). The budget stands alone in the example to avoid the direct relationship of changes in sales and production activities.

2 Capital expenditure budgets involve relatively large items of expenditure from the company. These investment expenditures are usually to improve the investment to improve future competitiveness. However, decisions would have to be made on existing asset sales and the raising of finance externally through loans or shares. Therefore, there is a high level of seniority involved in determining this particular budget. The authorized expenditures would have been proposed by functional heads and approved by the chief engineer, executive or managing director.

This budget has some similarity to the R&D budget. It is treated as stand-alone because the benefits of the expenditure are not likely to accrue in the immediate budget period but do impact substantially on the short-term projected cash-flow statements and balance sheets. In addition, some businesses remove the R&D facility from the normal business operation budgeting process because if the budget were to look unfavourable then it is this facility that is often the most likely to make cuts in expenditure without any obvious short-term consequences. In a competitive market this could prove costly for the firm in the longer run. Increasingly, because of operational pressure on such budgets as R&D, the function is made into an independent company and contracts with the parent company.

Master budgets

Essentially the master budgets for the budgeting period comprise of three major components:

1 A projected cash-flow statement for the period;
2 A projected profit and loss account;
3 A projected balance sheet.

The reader will recognize familiar terms in these reports. The principles in their construction are identical to those stated in Chapter 6. In essence the differences are:

(a) These documents can be formatted in any way that is desired by the Company's management. They are not constrained by external disclosure requirements because they are internal documents;

(b) There can be a number of iterations and drafts of the budget until the appropriate authority is satisfied that it is a budget the whole organization can commit itself to;

(c) The process in b. above is made much easier with the use of computing facilities including simulation models;

(d) The description above would suggest a democratic and participatory approach to the budgeting process. It is important to recognize that the management style may have a significant approach on the budgeting process. If the organization is highly centralized, with an autocratic management, the budgeting process may well be a top-down process. These behavioural implications of the budgeting process are very important, and have an impact on the motivation of engineers and managers alike. The issue is explored further in the final chapter of the book.

The subsidiary budgets and eventual master budgets are coordinated by the budget controller. In addition, the whole process has to be timetabled and the appropriate communications take place between the various departmental or functional dependencies. Their role can often be rather more wide-reaching in the light of the conflicts that often arise between parties with vested interests that may conflict with the company's.

Case study 9.1: Allgon

Prof. C. Lever, a chemical engineer, developed a gas, Allgon, suitable for a major refrigeration manufacturer. The next budgeted year she intends to increase production because of plans to sell 10,000 cubic litres at £100 per cubic litre.

At the beginning of the budget period she anticipates a stock of 4,000 cubic litres. This is considered too high and she would like to halve her physical investment in stock by the end of the budget period. There are two gases, coded AZ and BY, requiring five units and three units respectively, to make one cubic litre of Allgon. One unit of AZ costs £1 per unit and one unit of BY costs £0.50 per unit from suppliers. At the beginning of the budget period there will be 1600 units of AZ and 9600 units of BY. Because of potential production increases she would like to increase both raw material stocks by 25 per cent by the end of the budget period. Two processes are required to produce the gas, A and B. Process A takes four hours with conversion costs (labour and variable overheads) of £5 per hour and process B takes two hours with conversion costs of £7 per hour. Budgeted fixed costs are £200,000. Assume constant prices from one year to the next for valuation purposes.

Required:
1 The sales budget
2 The production budget
3 Material usage budget
4 Material purchase budget in £
5 Conversion cost budget
6 Budgeted profit for one unit
7 Budgeted Profit and Loss account

Solution:
1 Sales budget
Volume × Price= Revenue
10,000 × £100   = £1,000,000

The budgeted sales volume will be one of the variables to be used in the calculation of the production budget. The budgeted sales revenue figure can now simultaneously be transferred to budgeted profit and loss account. The sales budget would then be broken down into monthly budgets, and a timing of cash flow estimated from the cash or credit sales which have been planned.

2 Production budget
In order to determine production requirements, three variables have to be identified for the following equation to hold. They are sales, opening stock and closing stock of finished goods:

Production = Sales – Opening stock + Closing stock
(units)

Or, the same expression expressed slightly differently:

Production = Sales +/– (change in stock over the period).
Where sales is 10,000 units, opening stock is 4,000 units and closing stock is halved to 2,000 units then:

Production = 10,000 – (2,000) = 8,000 units

Alternatively,

	Units
If opening stock of finished goods	4,000
and closing stock to fall by 50 per cent	2,000
Then, falling stock	2,000
Because sales are	10,000
Production required	8,000

Stock has fallen and is therefore subtracted from sales.
If the stock position falls over the period then production will be less than sales because the shortfall will be met from the depletion of stocks.

The expression is similar to that employed in the determination of gross profit in Chapter 8 using the matching principle of:

Sales (units) = opening stock + production − closing stock

For profit and loss purposes we have identified the physical variables but this will have to be quantified in cost terms.

3 Material usage budget

From the production budget there is an output of 8,000 units to be met. In order to meet this output the inputs will need to be considered in turn. If the product is a standard product then the material requirements will be the same for the nth item as the first.

The material usage budget requires inputs of five units of AZ and three units of BY to produce one cubic litre of Allgon.

If budgeted production is 8,000 units then:

material AZ required is 5 units × 8,000 = 40,000 units
material BY required is 3 units × 8,000 = 24,000 units

The sums calculated are material requirements to meet production requirements. The amounts that have to be purchased will depend upon the existing stock position of raw materials.

4 The material purchases budget

This requires budgeted information from buyers concerning contract prices and terms. In a small business where buying may not be a separate activity, then the role could be undertaken by technical, production and engineering personnel. In the example, there are opening stocks that over time will need to be increased by 25 per cent at the end of the budgeted period. The budgeted purchase price per unit was given.

Hence, there is a similar equation that recognizes the buffer between purchasing and storage with production:

$$\text{Purchase} = \text{opening stock} + \text{usage of} - \text{closing stock}$$
$$\text{of materials} \quad \text{materials} \quad \text{of materials}$$

	AZ	BY
Opening stock	16,000	9,600
Closing stock	20,000	12,000
Increase of 25%	4,000	2,400
Materials required	40,000	24,000 (from previous budget)
Material purchase	44,000	26,400
If purchase price	£1	£0.50 (unit)
Then, cost	£22,000	£13,200

5 Conversion cost budget

As there are no stocks of variable labour and variable overheads then the costs incurred will only need to be based on the budgeted production output of 8,000 units. The units are expressed in hours:

Department A: 8,000 units × 4 hours × £5/hour = £160,000
Department B: 8,000 units × 2 hours × £7/hour = £112,000

Total conversion costs £272,000

6 A cost statement

A cost statement for one cubic litre of Allgon can be derived from the budget information:

	£	£
Material AZ (5 units @ £1)	5	
Material BY (3 units @ £0.05)	1.50	
		6.50
Dept A conversion (4 hours × £5)	20	
Dept B conversion (2 hours × £7)	14	
		34.00
Total variable costs		40.50
If selling price was		100.00
Contribution/unit		59.50

7 Profit and Loss statement

Finally, for C. Lever, there is sufficient information to produce a budgeted profit and loss statement:

Budgeted Profit and Loss account for the period

	£	£
Budgeted sales (10,000 units)		1,000,000
Budgeted variable cost of sales:		
Opening stock		
of finished product (4000 × £40.50)	162,000	

Add production (8,000 × £40.50)	324,000	
Deduct closing stock		
of finished product (2,000 × £40.50)	(81,000)	
		405,000
Contribution	595,000	
Budgeted fixed costs (sales, admin)	200,000	
Net profit		395,000

Check:
Produced in the budget year:

	£
Material AZ used 40,000 × £1	40,000
Material BY used 24,000 × £0.50	12,000
Conversion costs	272,000
Stock used over the budget year:	
2000 units @ £40.50	81,000
	405,000

Notes:
1 Stocks are valued at marginal costs and the budget accounts presented in marginal cost format (acceptable for internal reporting).
2 The marginal cost per unit has not altered because of price changes from the last period to the next.

Activity
Using the above example, identify additional financial information that would be required to provide a complete set of master budgets?
Re-cast the budget PLA in absorption cost format.

The importance of the cash budget

The previous example did not provide a complete master budget because of the limitations of the data in the case, i.e. there needed to be information normally provided in the balance sheet such as fixed asset purchases and disposals, creditors and debtors, capital introduced and repaid. Consider the following simple example:

Copy Katz, ex sales engineer, intended to begin his own business on the 1st July 19X3 with £20,000 capital in order to trade as an approved dealer of Taiwanese photocopiers. Each machine will be sold for £200 each and cost £120 to purchase. He had forecast the following sales demand:

	Units		Units
July	60	October	90
August	70	November	100
September	80	December	100

The nature of the business would involve extending sixty days credit. Payments are due for the supplies one month after delivery. At the beginning of July 250 photocopiers will be purchased on credit and in August 350. He would need to pay wages and other expenses of £4,000 per month. In addition, rent and business rates would amount to £4,000 for the six month period where £2,000 will be paid in July and the remainder paid in January the following year. Capital expenditure for a fork-lift truck will be required in January costing £6,000, with an expected life of three years and no residual value.

In order to establish Katz's financial needs for the six-month period, a budget statement of cash-flow, profits and balance sheet was required.

In order to answer this it would appear a number of subsidiary budgets are required. An important outcome of the subsidiary budget is the cash-flow budget. All of the other budgets, in this simple case, can be derived from the cash budget.

The format for a cash flow budget will vary. There are no set formats as this is an internal document. But it is important to ensure that it is clear, concise and easy to understand. A suggested format is:

Copy Katz budgeted cash flow for the six month period 19X3

£000's

	July	Aug.	Sept.	Oct.	Nov.	Dec.
Cash inflows						
Capital	20					
Sales			12	14	16	18*
1 Total inflows	20		12	14	16	18
less:						
Cash outflows						
Purchases		30	42			
Wages and other	4	4	4	4	4	4
Fixed assets	6					
Rent and Rates	2					**
2 Total outflows	12	34	46	4	4	4
3 Net cash flows (1–2)	8	(34)	(34)	10	12	14
4 Opening balance from previous period	–	8	(26)	(60)	(50)	(38)***
5 Closing balance (3 + 4 for next period)	8	(26)	(60)	(50)	(38)	(24)

* There were two further months of sales totalling £40,000. They would be included for profit calculation purposes but *not* cash-flow because they were sold on credit and not due to be paid by debtors until after the budget period.

** The rent and rates is an example of accruals for profit and loss purposes and should therefore include the amount owing. From a cash-flow perspective only £2,000 was to be paid in the period.
***The opening balance of the period is the previous period's closing balance.

Once the cash flow budget has been determined, then using both the data contained therein and making adjustments for stocks, depreciation and accruals (as in Chapter 6), the budgeted profit and loss and balance sheet for the period can be determined.

Budgeted Profit and Loss Account for the six months

	£000's	£000's
Sales (500 × £200)		100
Cost of sales:		
Opening stock	–	
+ Purchases	72	
– Closing stock (100 × £120)	(12)	
	60	
Gross profit		40
Wages and other (6 months × £4,000)	24	
Rent and rates	4	
Depreciation (£2,000 per 1/2 annum)	1	
		29
Budgeted net profit		11

Budgeted balance sheet as at 31 December 19X3

	£		£
Capital	20,000	Fixed assets	5,000
add Profit	11,000		
Bank overdraft	24,000	Debtors	40,000
Accruals	2,000	Stock	12,000
	57,000		57,000

The examples shown here used fixed budgeting. Fixed budgets are single point estimates of a particular level of activity or volume over the budgeted period. It is not adjusted to suit any other planned level of activity. Thus the fixed budget has some limited use at the first draft planning stage. However, if the subsequent actual levels of activity differ from the fixed budget it will have very little use for management control. The following section on flexible budgeting attempts to overcome this limitation.

Reconciliation of profit and cash

The budgeted cash-flow is negative £24,000 yet the profit is positive £11,000. There is a discrepancy of £35,000 between them. The profit was caused by anticipating revenue of £40,000 in the calculation. On the other hand, the cash flow did not reflect the full cost of rent and rates and the depreciation for the period:

	£
Profit	11,000
– Debtors owing	(40,000)
+ Rent and rates accrual	4,000
+ Depreciation	1,000
= Cash flow	(24,000)

This is fundamental to the understanding of profit measurement and cash-flow. From a planning perspective the cash deficit budgeted would have to be underwritten during the course of business – hence it would be appropriate to arrange overdraft facilities in advance from a bank. It should be pointed out that as a planning exercise the cash overdraft required over the planning period would be £60,000 in September, not the final £24,000. Thus unless this could be sustained further iterations on the budget would be required to avoid an unacceptable level of overdraft. In practice, the problem can be avoided by the following:

- Improved stock control and purchasing;
- Greater incentives to pay cash such as price and cash discounts (terms of trade);
- Delayed payment of purchases;
- Greater capital provisions in the form of cash;
- Better pricing;
- Cash cost reductions, e.g. wages etc.

This demonstrates the importance of business and project planning for the engineer and manager. The dangers of not recognizing potential cash deficits in advance should be obvious given the 1990s trading climate.

Electronic spreadsheeting

Budgeted information for any commercial financial planning purposes can be usefully modelled using spreadsheeting packages. It is particularly useful for asking 'what-if' questions through altering variables to determine the effects on the bottom-line of profitability or cash-flows.

Because spreadsheet formulas refer to the cell location rather than the numbers themselves, the formulas do not have to be altered to accommodate any proposed changes in the numbers themselves.

Flexible Budgeting

Flexible budgeting is a related tool that lends itself to both planning and controlling, particularly the latter.

The model identified in Figure 9.2 is an example of a fixed budget where the output and activity levels, once arrived at, are fixed for the period. This means that if there are activity changes from the fixed budget intended then it is probable there will be substantial differences on nearly every cost and/or revenue item, with the possible exception of fixed costs.

The financial differences between budget and actual are commonly known as *variances*:

Budget £ compared with Actual £ = Variance £

For example, sales and production of Orion, a standardized agricultural machine, were fixed at 10,000 units for the budget period. The budgeted selling price was £200 per machine and the budgeted variable cost of each machine was £150 for the same period. The actual sales were 8,000 units, with actual revenue being £1.68 million and actual costs being £1.36 million.

Using a fixed budget approach to identify the differences between what should have happened and what did actually happen will reveal the following:

			Fixed budget £m	*Actual £m*	*Variance £m*
Sales revenue	10,000 × £200	=	2	1.68	0.32m (A)
Variable costs	10,000 × £150	=	1.5	1.36	0.14m (F)
Contribution	10,000 × £50	=	0.5	0.32	0.18 (A)

The budgeted contribution should have been £500,000. The actual contribution was £320,000. In a competitive environment this could cause the firm greater problems by the end of the financial reporting year unless managers take some action in marketing, sales, production and engineering decisions.

If at the end of the period the profits were less than budgeted then the role of management ought to be to examine why the profits were less and to take action where needed. This means examining in detail what contributed to the shortfall in profit. Identifying where the major responsibilities are will require a more detailed analysis.

For control purposes there is sufficient information to enable variable cost control to be undertaken as it appears there is a favourable variance of £0.14m. Although profitability is less than that budgeted, it is understandable if the level of activity of production and sales had fallen. A further question that needs to be raised is: given the actual output levels that were achieved, how efficient was the business using its resources?

In order to answer the question the budget needs to be revised and show the budget figures that should have been achieved with the benefit of hindsight on the actual level of activity:

	Units 10,000	8,000	8,000	
	Fixed £m	*Flexible £m*	*Actual £m*	*Variance £m*
Sales revenue	2	1.6	1.68	0.32 (A)
Variable costs	1.5	1.2	1.36	0.14 (F)
Contribution	0.5	0.4	0.32	0.18 (A)

The flexible budget is the original budgeted unit data multiplied by the actual output. This enables greater information to be provided in financial terms to those controlling both costs and revenues.

When the fixed budget is compared with the flexible budget there are *volume* variances. These variances are, in the main, the responsibility of senior engineering, marketing, production and sales management.

When the flexed budget is compared with the actual the variances are known as price and expenditure variances:

£m	Volume	Price/ expenditure	Total
Sales	0.4 (A)	0.08 (F)	0.32 (A)
Variable costs	0.3 (F)	0.16 (A)	0.14 (F)
Contribution	0.1 (A)	0.08 (A)	0.18 (A)

There should have been a fall of £300,000 in variable costs because of the volume fall. The total fall was only £140,000 because of an adverse variance in expenditure of 0.16m (A). This information will give different signals to engineers and management in exercising their control function than an overall favourable £0.14 (F).

Activity and volume levels are often the responsibility of relatively few key decision-makers in the organization. Throughout the organization engineers and managers are responsible for the use of resources. They will need only to be given the information that they have responsibility over.

The format used in this exercise was marginal costing. If absorbed costs had been used the fixed costs of production and sales would need to be identified separately. In general, fixed costs would not need to be flexed because they do not respond to changes in output over the short-term.

The above expenditure variances can be broken down into further sub-variances. Greater explanation on the principles of variance analysis will be discussed in the next section under standard costing.

Standard costing

A standard cost is a predetermined cost of a product. It is particularly appropriate for large scale discrete (batches) or continuous (process) manufactured output. For example, if a newly designed motor car is to be manufactured in large numbers from a prototype, a standard product specification would be drawn up by engineers. The specification would identify, for each stage of production, the inputs required for each car under specified conditions. The inputs identified would include:

- raw materials required
- components and sub-assemblies
- direct labour
- variable overheads

Each item of resource would be costed in advance using, for example, the procurement department's knowledge and skills in purchasing from suppliers, the personnel department's proposed rates of pay for particular job grades under a specified set of conditions. Industrial engineers and work study would play a major role in determining the production department's productivity, machine speeds, set-up and tooling change-over times.

It would not be necessary for all these specifications and costs to be newly generated for each new project or product, since the new car is likely to use techniques, processes and components used in the firm's existing product lines, and some of the specifications will be similar or identical to those already in existence.

The final specification could be kept on record using a computing database. The specification and standard cost will be maintained and updated by authorized personnel according to engineering and product developments and other significant changes affecting the product. Periodically there would need to be a complete review of the whole specification and standard cost.

Standard costing is a technique that complements budgeting systems for planning and control purposes, by basing budgeting decision on these standard costs. At the end of a product life the specification would be withdrawn.

A standard cost has some essential features:

1 it is an *engineered* cost
2 for a *standard* product
3 on a *cost/unit* output basis.

Industries with firms employing standard costing are bricks, cement, food processors, detergents, chemicals, cars, photocopiers, micro computers etc. Increasingly, the service sector is making use of standard costing, including conveyancing, building society accounts etc.

It was stated earlier that the use of standard costing is compatible with the budgeting exercise. Their major similarity is in the ability to assist planning and control.

Budgets are concerned with cost centres, revenues, departments, divisions. They involve allocated total resources expressed financially. Standard costs, on the other hand, are the inputs of resource used to produce one unit of output.

Their relationship can be summed up in the expression:

Budget costs for a period = Budgeted volume production
× Standard cost per unit.

There is no legal requirement for a firm to employ standard costs. On the contrary there are considerable expenditures involved in setting-up and maintaining a standard costing system.

The benefits of standard costing

1 It assists budgetary control and responsibility accounting through the provision of *variances*.
Variances: Periodic comparisons expressed financially between budget totals and actual.

i.e. £ Budget v £ Actual = £ Variance

Actual: Variances need to be reported in a timely and relevant manner to the engineers and managers responsible for resources.

The financial recording system explained in Chapter 6 can provide this information. Each variance produced should be compatible with an individual engineer's or manager's resource responsibilities.

2 The predetermined nature of a standard cost enables businesses to consider revised pricing policies, discounts, and terms of trade for their products through revised price lists in advance of the trading period.

3 Internal periodic profit measurement can be undertaken with greater frequency internally by valuing the stocks at a standard cost. This avoids both a reliance on physical stock-taking and accounting oriented approaches such as FIFO and LIFO.

At the year end there would need to be a reconciliation of stock value for profit measurement for external purposes and some physical stock-taking for audit purposes. Standard costing stock values would not be acceptable for the final external accounts but would be acceptable for periodic internal management accounts.

4 Variances have additive properties. These should conform to the increased responsibilities of managers further up the hierarchy as the information's level of detail reduces. This is commonly known as responsibility accounting.

Any given product will have its own complexities which are bound to produce variances – adverse and favourable. It is not the responsibility of engineers and managers to investigate every variance but only those that could be regarded as significant – both in absolute and statistical terms.

Weaknesses of standard costing

1 Standard costs need to be maintained regularly and reviewed. This can be mistakenly regarded as costly and time-consuming, causing inaccuracies and figures that are meaningless for planning and control purposes.

2 There needs to be a review of the timeliness and relevance of variances provided by the accounting system. If these conditions are not satisfied then confidence in the control system can be undermined.

A standard cost can be presented in either an absorbed cost or marginal costing format. The latter method is often more user friendly for non-accounting managers.

Variances

Earlier it was stated that a variance is a difference between *budget* and *actual*. For a whole business at the end of a period there is one total variance:

Budgeted profit v Actual profit = Profit variance

If, however,

Budgeted profit > Actual profit = Adverse profit variance

then, this would be considered as an adverse (A) or unfavourable variance to the business because profits were not as high as were anticipated.

The opposite is the case when actual profit > budgeted profit. This is known as a favourable (F) variance.

The overall responsibility of this variance is likely to lie with the senior executives of the organization, such as managing directors. However, while taking the overall responsibility, the ability to control the variance can only be carried out by other engineers and managers within the organization. They have specific responsibilities for various costs and revenues contributing to the overall profit measurement for the period.

The term 'cost' is simply a financial measure of the resources used. In a business, costs measure resources that have been used in undertaking purposeful activity directly with output or, to support output. The

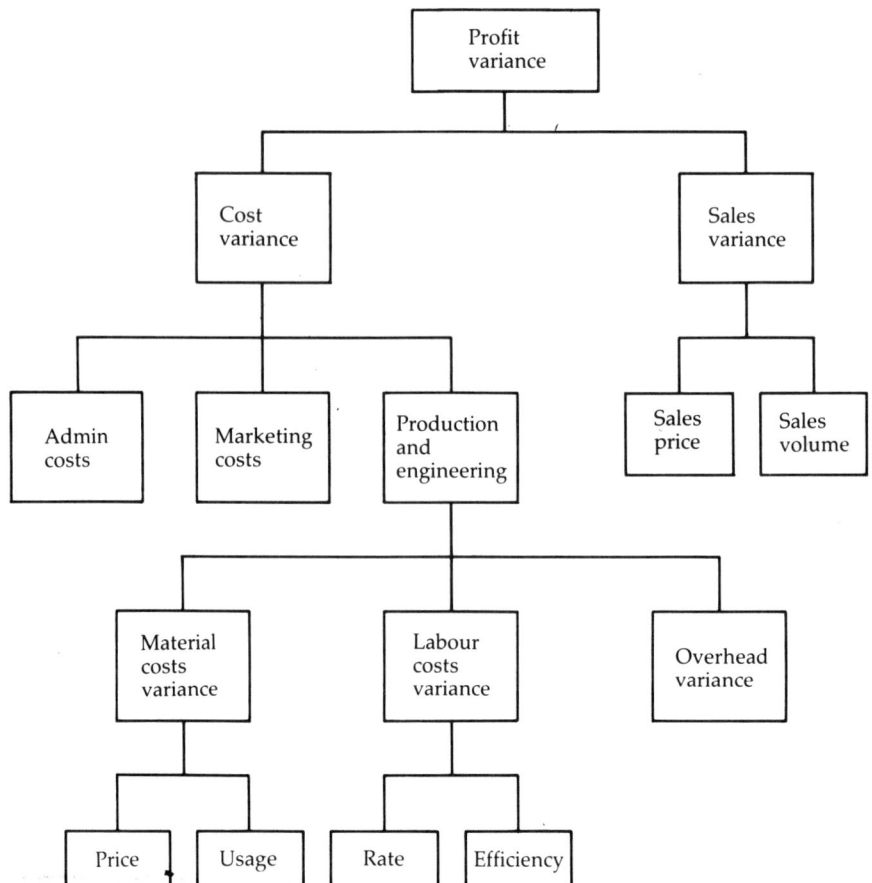

Figure 9.3 Cascade effect of variances

purchase of items from retail outlets often involves a cost to the purchaser. Buying one shirt at a price of £15 involves a cost to the purchaser of £15. The receiver, in this case the retail outlet, charges a price for its goods. (Price is often used interchangeably with cost when there is only one of an item.)

Price is simply the amount charged by a provider of a unit of resource to a customer or client.

Therefore price and cost are terms that are used interchangeably depending on where the business is in the exchange process of buying and selling.

If selling: price
If buying, acquiring or using: cost

Any cost is a function of two variables:

purchase price × quantity of item purchased of the input.

For materials, there will be a quantity of specified materials and a price per unit of material be it kilograms, barrels or boxes.

For labour, there will be a quantity of specified employees. Because of different tasks and skill levels labour is often expressed in terms of hours. The rate of pay is traditionally expressed in terms of £/hour depending on job grades and skill levels etc. Salaried and professional employees can be expressed in terms of man-hours.

Calculation of variances

In flexible budgeting it was shown that expenditure variances could be determined by comparing the cost of actual output with the budgeted cost of that same output. Therefore, after a measured output has been achieved, the question is one of how much items *did* cost in obtaining the output and, how much they *should* have cost in achieving the same output. In essence the emphasis is firmly placed on *control*.

To make one standard unit of a product (output) a standard amount of materials input will be used at a standard price per unit (input):

Standard quantity of input × standard price = standard cost/unit.

For example, if twenty kilos of material Y were required at a standard price of input of £8 then the standard cost of input to make one unit of output would be:

Standard Quantity (SQ) × Standard Price (SP) = Standard Cost

20 kilos × £8 per kilo = £160

From the recording system, the total actual cost of the input will be dependent on the actual output. If 100 units of output used 1900 kilos of input of material Y at a cost of £17,100 then there is sufficient information to determine the cost variance:

	£
100 units should have cost 100 × £8 × 20 kilos =	16,000
100 units actually cost	= 17,100
Cost variance	1,100 (A)

The variance on the cost of material is adverse because the costs incurred were greater than anticipated. Those responsible for controlling the cost would be expected to investigate. However, the cost is a composition of two elements that might entail separate responsibilities in a large organization. Thus a breakdown of this variance is needed.

The material price variance will require the standard price of input to be compared with the actual price of input and the difference multiplied by the actual quantity used. The standard price is given. The actual price is £17,100/1,900 kilos = £9.

Thus: (standard price – actual price) × actual quantity
= (SP – AP) × AQ
= (£8 – £9) × 1,900 kilos = £1,900 (A).

The price variance is adverse because the price paid was greater than anticipated according to the standard set.

The material usage variance will require the standard amount of input to be used in achieving the actual output to be compared with the actual quantity used and then multiplied by the standard price.

Thus: (standard quantity in actual output – actual quantity used)
× standard price of input
= (SQ – AQ) × SP
= ((20 kilos × 100 units) – 1900 kilos) × £8 = £800 (F)

The usage variance is favourable because the quantity used was less than what should have been used in obtaining the actual output.

In summary the variance of material Y is:

	£
Price variance	1,900 (A)
Usage variance	800 (F)
Cost variance Y	1,100 (A)

Moreover, the significance of calculating the sub-variances enables those responsible to control their element of the cost.

The purchase price is the responsibility of procurement and the usage variance is the responsibility of those using the material i.e. production and/or engineering. Of course, there will often be some degree of interdependence, for example, if there were favourable price variances and adverse usage variances there might be an issue of differing quality of materials due to a change in the supplier.

In reality the composition of the product will rely on a number of different materials which should allow for standard wastage, evaporation, distillation in process etc. Nevertheless, the same principles should apply.

Furthermore, if the standard cost of labour can be determined then the same principles are applied to the control of labour. The terminology changes because labour usage will be measured in hours and the price will be measured as a rate per hour. There will be an overall labour cost variance broken down into sub-variances of rate and efficiency respectively.

Worked example of variances

The Zantan Co. has adopted the following budget for the year for a standard value using standard costs:

Production	20,000 units
Direct materials	20,000 kg at £2/kg
Direct labour	15,000 hours at £8/hr
Variable production overheads	£90,000
Fixed production	£120,000

After one control period (one month), actual information was provided by the production and engineering departments:

Production	1,500 units
Direct materials	1,800 kg costing £3,420
Direct labour	1,300 hours costing £11,050
Variable production overheads	£8,450
Fixed production	£11,400

If the original budgeted information is flexed then:

	Flex 1,500 units £	Actual 1,500 units £	Variance £
Direct materials (1,500 × £2)	3,000	3,420	420 (A)
Direct labour (1,125 × £8)	9,000	11,050	2,050 (A)
Variable production overheads	6,750	8,450	1,700 (A)
Fixed production	10,000	11,400	1,400 (A)

The flexed budget has been obtained by flexing the variable costs with output using the standard costs: direct materials, direct labour and variable overheads. The fixed costs do not alter with output and have simply been arrived at by dividing the budget year into twelve equal months.

There are four adverse variances that need to be examined by engineers and managers for control purposes.

Direct materials

(Actual price − Standard price) × Actual quantity
((£3,420/1,800) − £2) × 1,800 = £180 (F) Price variance
(Actual usage − Standard usage in actual output) × Standard price
(1,800 − 1,500) × £2 = £600 (A) Usage variance

The overall material cost variance is £420 adverse.

Direct labour

Same principles as materials except rate is substitute for price and usage for hours:

(Actual rate − Standard rate) × Actual hours
(£8.50 − £8.00) × 1,300 hours = £650 (A) Rate variance
(Actual hours − Standard hours in actual output) × Standard rate
(1,300 − (15,000/20,000 × 1,500)) × £8 = £1,400 (A) Efficiency variance.

The overall labour cost variance is £2,050 (A).

Variable overheads

Because the incidence of variable overheads is caused by labour which is treated as variable, the variable overhead variance will be calculated the same as the labour variances in this case

(Actual rate – Standard rate) × Actual hours
(£6.50 – £6.00) × 1,300 = £650 (A) Expenditure variance

(Actual hours – Standard hours in actual output) × Standard rate
(1,300 – 1,125) × £6.00 = £1,050 (A) Efficiency variance
The overall variable overhead cost variance is £1,700 (A).

Fixed overheads

In this case the budgeted fixed overheads for the period are compared with the actual fixed costs for the same period:

Actual fixed costs – Budgeted fixed costs
£11,400 – £10,000 = £1,400 (A) Expenditure variance

All four major cost variances have now been broken down into their respective sub-variances where possible, allowing management control to take place for those responsible for their constituent element of costs.

Significance of variances

The significance of variances would need to be identified in advance for control purposes otherwise there would be an information overload rendering the most important variances liable to be overlooked. If a Pareto approach was taken, for example, 20 per cent of variances would account for 80 per cent of the costs, of which some would be statistically significant and require management attention. This concept is known as 'management by exception'. It would also be advantageous to keep a cumulative check on variances besides individual time periods to assist the control process.

Rolling budgets

The budgeting process given in the example appears to be fixed for one year in advance. Any control exercise is through comparing actual financial results against the budget broken down into control periods. The rolling budget enables a continuous twelve-month horizon. The results in the first monthly control period may allow the future twelve-month budget to be modified and include another control period simultaneously. Such a system needs to be carefully managed and requires sound accounting data gathering and strong communication links between engineers and managers.

Zero base budgeting

This involves an annual justification of all financial resources used by managers, department or function. For example, if the output of an engineering design team had declined, the budget would reflect this. If this continues then indeed its whole future as a function might be questioned by management in terms of future resource commitment.

Chapter summary

The larger and more complex the organization the greater the need for an effective budgeting system. Budgeting is the planned financial 'blue-print' for the future short-term planning horizon of the business organization. It is intended to plan and control the acquisition and allocation of resources. The budget conforms to the systematic engineering and management activities of planning, organizing, monitoring and controlling.

The chapter identified the overall interdependence of the budgeting process beginning with the compilation of functional budgets such as engineering and sales and ending with a summarized budgeted profit and loss account and balance sheet together with a budgeted cash-flow statement.

The fixed budget provided a reasonable framework for single-point planning purposes. In practice a flexible budget will identify a range of budgets based on differing levels of business activity. This not only enhances the planning function but enables responsibility accounting and control to take place by identifying variances between the budget and a given level of actual output. Further analysis of the variances for management control purposes could be undertaken by using the technique of standard costing.

Further reading

The Further reading references given at the end of Chapter 8 cover this material well, with the addition of Chadwick, *The Essence of Management Accounting* (Prentice Hall, 1991).

Tutorial exercises

9.1 Blew Films

Mr Blew is a sales rep. in the photographic film business and has been for many years. He has recently inherited £30,000 which he is prepared to put into a business venture, or anything else which could improve his income. He has an idea to go into business, selling specialist photographic films used in medical laboratories.

He reckons that he needs to buy £30,000 worth of stock to begin with, and £10,000 a month afterwards to replace sold items. He will settle his bills one month in arrears.

He expects to sell £10,000 worth of equipment in the first month, and £12,500 per month afterwards, at a margin of 20 per cent; meaning that one fifth of the sale price is gross profit. He expects his customers to pay one month in arrears.

He has a lease on a warehouse and equipment for £2,000 per year, paid quarterly in the first month of each quarter. Running costs of the warehouse should be £200 per month, paid one month in arrears.

He has found a suitable van, which he can buy for £7,500, and he estimates that it will depreciate by £2,000 in the first year (using the straight line approach). Running costs will be £150 per month, paid one month in arrears. Tax and insurance on the van will come to £250 per year, paid in advance at the time of purchase.

To do this work, he will have to give up his present job, from which he earns around £1,000 per month, before tax, national insurance etc. He thinks he can't manage on an income of less than £1,000 gross per month, drawn from the business.

(a) Draw up a forecast cash flow statement, and a forecast trading, profit and loss account and forecast balance sheet for the first six months trading.

(b) Would you advise Mr Blew to go into this business? If so, would you give him advice on, say, overdraft limits? What other advice would improve the chance of his business being a success?

9.2 Black Holes

Mr Black has been a self-employed construction engineer for many years. He has recently inherited £15,000 which he is prepared to put into a business venture, or anything else which could improve his income.

He has recently noticed that the quality of the holes on many golf courses is rather poor; the metal lining having rusted away over a few years. He believes he can import new Japanese holes, made from a resin which will not deteriorate in wet weather.

He reckons that he needs to buy £5,000 worth of stock to set him going, and a further £2,500 a month to replace sold items. He will settle his bills one month in arrears.

He expects to sell £5,000 per month, at a margin of 50 per cent; meaning that half of the sale price is gross profit. He expects his customers to pay up one month in arrears.

He has a lease on a storeroom for £2,000 per year, paid quarterly in the first month of each quarter. Running costs of the storeroom should be £25 per month, paid one month in arrears.

He has found a suitable van, which he can buy for £12,000, and he estimates that it will depreciate by £3,000 in the first year (using the straight line approach). Running costs will be £400 per month, paid one month in arrears. Tax and insurance on the van will come to £220 per year, paid in advance.

To do this work, he will have to give up some of his engineering activities, from which he earns around £2,000 per month, before tax, national insurance etc. He reckons he would have to give up around half of his present work, so will need to make drawings of around £1,000 per month.

(a) Draw up a forecast cash flow statement, and a forecast trading, profit and loss account and forecast balance sheet for the first six months.

(b) Would you advise Mr Black to go into this business? If so, would you give him any advice on, say, overdraft limits?

9.3 Green Grass and Son

Mr Green has been a self-employed landscape gardener for many years. He has recently inherited £30,000 which he is prepared to put into a business venture, or anything else which could improve his income. He has an idea to go into business, selling specialist grass seed and rare conservatory plants to the up-market gardening firms in the country.

He reckons that he needs to buy £30,000 worth of stock to set him going, and a further £7,500 a month to replace sold items. He will settle his bills one month in arrears.

He expects to sell £10,000 per month, at a margin of 25 per cent; meaning that a quarter of the sale price is gross profit. He expects his customers to pay up one month in arrears.

He has a lease on a warehouse and equipment for £8,000 per year, paid quarterly in the first month of each quarter. Running costs of the warehouse should be £200 per month, paid one month in arrears.

He has found a suitable van, which he can buy for £7,500, and he estimates that it will depreciate by £2,000 in the first year (using the straight line approach). Running costs will be £400 per month, paid one month in arrears. Tax and insurance on the van will come to £750 per year, paid in advance.

To do this work, he will have to give up his present gardening activities, from which he earns around £1,000 per month, before tax, national insurance etc. He thinks he can't manage on an income of less than £12,000 gross per year.

(a) Draw up a forecast cash flow statement, and a forecast trading, profit and loss account and balance sheet for the first six months.

(b) Would you advise Mr Green to go into this business? If not, what steps could be taken to make the business more attractive (incorporate the effects of any change in to the budgeted accounts)? What advice would you give on overdraft limits if he were to start up?

9.4 Loopy Loos

Ms Loopy has been a self-employed plumber for many years. She has £2,000, which she is prepared to put into a business venture, or anything else which could improve her income.

A friend of hers runs a small manufacturing company, specializing in adhesive labels and transfers. They have worked with a designer, to produce a range of stickers, which can be stuck on lavatory pans, making the toilet resemble well known political and media figures. Ms Loopy thinks there would be a large market for these, reached through joke-shops and gift shops, and would be prepared to become a self-employed saleswomen for the products.

Neither the manufacturer nor designer are interested in a joint venture or partnership, but the designer will accept a fee of 5 per cent of gross sales, paid quarterly, one month in arrears (i.e. the royalty payment for the first three months will be paid in month four etc.). The manufacturer is prepared to offer one month's credit on stickers purchased.

Ms Loopy reckons that she needs to buy £15,000 worth of stock to set her going, and further purchases of £8,000, £10,400, £12,800, £15,200 and £16,800 respectively for the next five months.

She will start by contacting suitable outlets in her own county, and expects to receive £10,000 in sales receipts per month for the first two months, at a margin of 20 per cent; meaning that 20 per cent of the sale price is gross profit. She expects her customers to pay up one month in arrears. Afterwards, she expects sales to increase by £3,000 per month, for the next six months (meaning that sales in month three and four are £13,000 and £16,000 respectively) as she makes contact with outlets outside her immediate locality, but doubts that the margin will improve by much.

She will use a spare room in her house to co-ordinate her sales activities. She estimates that running costs of the business (mostly postage and telephone expenses) will be around £120 per month. She will pay these bills one month in arrears.

She will buy a suitable van, currently valued at £4,500, and she estimates that it will depreciate by £500 in the first year (using the straight line approach). Running costs will be £350 per month, paid one month in arrears. Tax and insurance on the van will come to £180 per year, paid in advance. She will also have to pay £150 to have the slogan 'Loopy Loos' painted on it.

To do this work, she will have to give up all of her plumbing activities, from which she earns around £800 per month, before tax, national insurance etc., but can't live on less than £800 per month to cover food, rental on the house etc.

(a) Draw up a forecast cash flow statement, a forecast trading, profit and loss account and a forecast balance sheet for the first six months (or any other time period if you think it more helpful). Ignore any tax advantage that might result from using part of her rented house and property as a business.

(b) Would you advise Ms Loopy to go into this business? If so, would you give her any advice on, say, overdraft limits?

(c) Can the main problem facing the business in its first year be overcome in any way? Incorporate your suggestions into the budget.

Case study 9.2: Ascom

Ascom puts lack of control as reason for loss

By Ian Rodger in Zurich

Ascom, the Swiss telecommunications equipment group, blamed inadequate internal controls for the belated discovery last month of SFr61m ($41m) in extraordinary losses at its German subsidiaries.

The discovery forced the group to announce last week that it would suffer a 1992 net loss of SFr46m and pass its dividend rather than achieve the roughly SFr25m net profit forecast a few days earlier in a letter to shareholders.

Mr Leonardo Vannotti, president, accepted the full blame for the debacle, which has caused Ascom shares to lose about 14 per cent of their value.

However, he said at the group's annual press conference yesterday that no member of group management would be resigning.

The losses arose in the group's cable television and mobile radio subsidiaries in Germany. Its cable franchises, purchased only three years ago, have required heavy investment in infrastructure, but so far have attracted few subscribers. The mobile radio business was left with large stocks of technologically outdated products.

In both cases, internal reporting procedures were inadequate and it was only after line managers were replaced in March that the need for large write-offs was discovered by external auditors.

Mr Vannotti said group management had been working hard on improving internal controls at Ascom but it was difficult to find the right people for these positions.

The group, the product of a 1987 merger of three traditional protected suppliers to the Swiss PTT, had not needed strong controls in the past.

Both Mr Vannotti and the board believed that the group's new strategy to expand in specialized global niches was still the right one and they were confident that the recent reporting failure would not recur. It was also clear that group managers had not deliberately misled investors.

Therefore, there was no point in any of them resigning.

Mr Vannotti said he expected the group to return to profit this year.

Financial Times, May 1993

9.5 What budgeting and control issues are raised by the case above?

10 Production management and manufacturing strategy

Whereas manufacturing management is primarily concerned with the physical processing of goods, production management is about the efficient and effective use of labour, materials and overheads matched to market demand for the products with which the company serves its customers.

This chapter draws on earlier marketing, production and accounting chapters to place manufacturing into a strategic context. It examines how the day-to-day running of production systems, be they producing engineered components, food packaging or plastic mouldings etc., can make best use of expensive human resources, machinery, premises and working capital to operate the business and add value for the consumer. Production decisions will be examined by looking at their effect on the balance sheet, manufacturing account, trading profit and loss account and cash flow. First, we examine the effects of production management on costs of production and the balance sheet, in the short and medium term, and then outline the longer term strategic implications of manufacturing strategy.

Production management

Production control will influence working capital and cost of goods sold, both measures for financial effectiveness and reported in the balance sheet and profit and loss account (PLA) respectively.

Working capital and production

Working capital was discussed extensively in Chapter 7. It is the lifeblood needed to run any business, measured by taking snapshots of where the money is 'tied up' at a particular moment in time. This is done by regular monitoring of the balance sheet.

Raw materials

Precisely what it implies: examples such as sheets of metal, plastic and leather, or rolls of electrical wiring in their uncut form: in other words the materials needed at the 'pre-operations stage' prior to cutting, machining, tanning and other general preparation for the subsequent value adding processes which make up the company's core business.

Most companies recognize materials management as a crucial function in its own right, where important savings can be made through shrewd *sourcing*. This works best when purchasing and design engineers are both aware of their objectives: the former on getting the best deals available,

Figure 10.1 Working capital tied up on factory floor

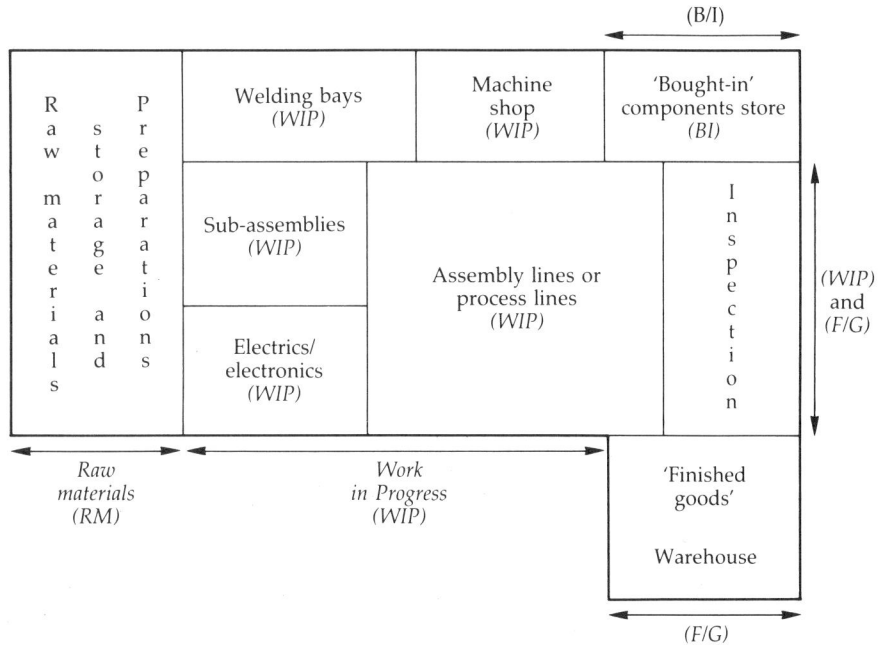

and the latter, in a prototype buying aspect, on specifying the optimum value for money on components for new, modified or improved products. Major cost savings can be made by designing for manufacture at the R&D stage.

Modern manufacturing businesses tend to sub-contract a significant proportion of component processing to component specialists, hence purchasing components ready for major sub-assembly and final assembly. However, 'raw materials' refers to whatever state materials and components are received by the company prior to its own value adding processes. The traditional philosophy of buffer components stocking as an insurance against wayward suppliers slowing down the production process, can lead to an unnecessary amount of raw material inventory being carried, which incurs storage costs and ties up cash unproductively.

A major objective of modern production planning and execution is control, to the point of virtual elimination, of raw material stock for which the company does not have committed and firm customers' orders, using JIT, or some hybrid system which minimizes stock level wherever and whenever possible.

Work in progress

This consists of labour input and associated production overheads involved in sub-assembly and major assembly processes – for example: a tiny value of each piece of machinery in the process is amortized to each unit as it goes through that machining process. This accountancy process is covered in other chapters, but the production process has to be controlled manually or electronically at all times, so that the amount of productive and non-productive labour can be accounted for, and allocated to, individual cost centres by the company's accountants.

Much can be done, in terms of factory floor layout and FMS to minimize the time taken to add value on the production line. When a firm can combine a fast throughput via FMS and JIT (where products are made to order, see Chapter 3) surprisingly little WIP will be located in the factory at any one time.

Finished goods

These are items duly processed, and ready for despatch. Control of finished goods is often a point of contention at management level, as to which function should control this customer orientated aspect. Production people are closest to the physical stock, whereas sales and marketing personnel should, by virtue of their customer proximity, have the accountability to dictate finished stock levels. Many companies still allocate this accountability to the works manager. The application of JIT makes it much easier to allocate this stock to where it belongs: namely with the customer; and the departments closest to the customer are sales and marketing. Stocks can often be reduced by 50 to 60 per cent, for example, for a £20 million business with £4 million stock, this would release capital of £2 to £2.4 million, for reinvestment or reduction in borrowing. Figure 10.3 relates to the case study below, and shows how these savings are achieved through the constituent parts of stock: i.e. raw materials, work in progress and finished goods. The chart shows value of each and how, typically, it would reduce over time.

The idea of inventory turns

The aggregate of all components, in raw, semi-finished or finished state, but not yet invoiced for delivery to customers, is called 'stock and work in progress', on the balance sheet, and is a measure of total inventory used to run the business. Production management, and cost control generally, are interested in and accountable for turning this inventory as rapidly as possible. For example a manufacturing business needs £½ million tied up in inventory at all times to achieve an annual revenue sales of £2 million.

The business therefore has an inventory turn of 4 per annum (revenue/inventory value). Production management will ask the question: why do we need £½ million working capital to generate £2 million when, for example, our competitors only need £¼ million? It may be that the competitor's inventory control is substantially better, perhaps based on JIT, where *finished inventory* is not produced until an actual order is received from the customer. On the other hand, it may be that the products or the technologies involved are significantly different; for example Morgan would be expected to have fewer inventory turns on their hand-built sports cars than would Lotus, under normal circumstances.

Cost of goods sold

From PLA we get the cost of traded goods, or cost of sales, broken down into:

- materials consumed: typically 45–65 per cent of cost of sales,
- direct labour: typically 8–20 per cent of cost of sales,
- production overheads: typically 20–45 per cent of cost of sales,

as part of the prime costs of production. Modern production methods, such as JIT, can do much to reduce the first two, although higher depreciation costs may lead to higher overhead costs in some cases. The importance of this in terms of cost control and competitive strategy can be examined through the following case study.

Case study 10.1: Electric Motors

Product Costs for Electric Motors

In the early 1980s, the market for small electric motors (used to power office machinery, for example) became very competitive. Figure 10.2 shows the cost structures that were developed by the three leading competitors, based in the UK, West Germany and Japan.

The major differences in cost are outlined below:

Materials	Highest in UK, largely because of small batch sizes. In Germany, batch sizes were twice that in the UK, but in Japan, batch sizes could be anything up to ten times larger. Consequently, discounts were more easily obtainable than in the UK, where wastage rates were also much higher.
Labour	The low wage costs in the UK were more than offset by higher productivity elsewhere.
Production overheads	Lowest in Germany and the UK, both using traditional production methods. Japan's costs are higher, largely due to the depreciation charge on the capital equipment used.
Profit margin	The three companies are shown at 20, 25 and 30 per cent respectively, although this is somewhat notional. As competition intensified, the UK based company felt it had done well if it achieved 10 per cent on new business.

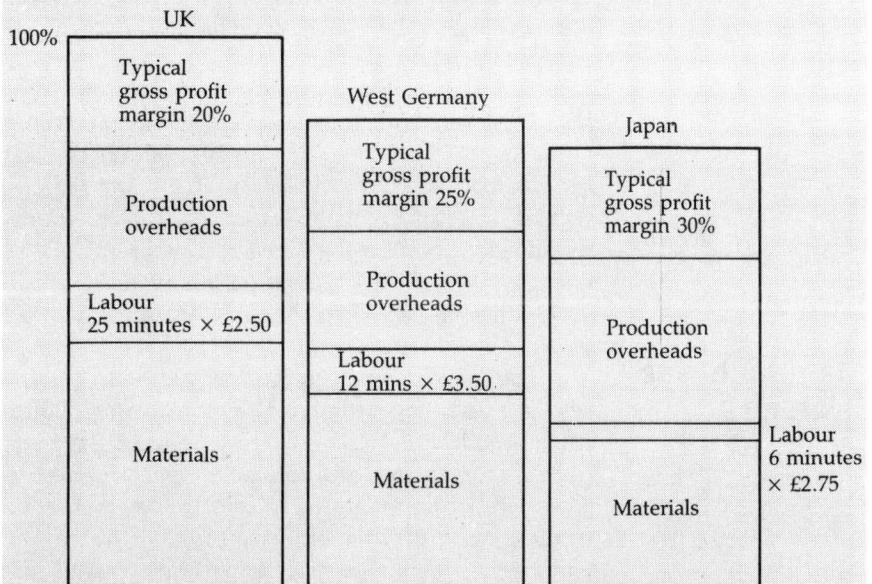

Figure 10.2 Cost structures in an electrical motors market, mid-1980s

> The UK company found itself in increasing difficulties and eventually it was divested. The buyer was able to make a success of it by integrating the product range into its own Japanese style production system.

Production, manufacturing and the business environment

It is tempting to think of manufacturing and production as part of the internal business environment, and overlook the contribution they can make to building robustness to adverse changes in the outside business environment, and the direct contribution both can make to developing competitive advantage.

Production control and the business environment

Modern manufacturing methods, and JIT in particular, can do much to reduce the working capital and cash flow problems experienced by UK industrialists in the recessions of the 1980s and 1990s. As consumer purchases fell during these recessions, production levels also fell. The knock-on effect of this caused cash flow pressures on key suppliers. These cash flow shortfalls necessitated financing by extra borrowing (at record high interest rates). Banks became reluctant to lend to firms with falling order books, exacerbating the problem still further.

Hence, Western businesses, with no escape route via their customers as cash raisers, and certainly none from the banks during a recession, were forced to look at other means of arresting a cash flow crisis. Many companies 'de-stocked' by liquidating inventory without replacing it. But this was only the first stage of survival strategy, which many did not make. Redundancies and factory closures followed, with the subsequent lessons learnt about not creating further inventory unless absolutely needed.

Market pressures thus forced a radical rethink on minimizing the need for working capital. The survivors still had the challenge of evolving new working methods which involved much more effective use of working capital. Increasingly, the Japanese manufacturing philosophies methods of JIT, TQM and FMS took on credence in the West.

In Chapter 9 we looked at the breakdown of production costs into materials, labour and overheads, how these are dealt with in the PLA, and the need to record constantly and review how much material, labour and overheads are allocated to each unit produced. The following case study, Company Q, shows how a company used production control to reduce costs following adverse changes in the competitive environment.

Manufacturing strategy

The use of modern production methods alongside a strong, proactive research and development function can be combined to construct powerful competitive advantages. In this section we briefly outline the kinds of manufacturing strategy commonly encountered today, but it is important to remember that manufacturing strategy cannot be used to lead a company in the long term; it is simply an important way of generating a distinctive competence in the way a firm creates value for its customers. It is the combination of marketing and manufacturing strategy which creates a devastating advantage.

Case study 10.2: Company Q

Towards a production advantage

This study examines cost control issues, involving the vital interface between engineers, buyers and manufacturing engineers who plan the work processes.

Company Q had been a marketing-led producer of standardized engineering components for many years, and had always produced a good return on capital for its owners. Although its products were not particularly distinctive or cheap, the company's pre- and after-sales services were extremely good.

The first hint of trouble came in the mid-1980s, when the sales team reported that although the existing customer base remained very loyal, it was becoming increasing difficult to obtain new business. A marketing audit and competitive analysis showed that the company was being undercut by conglomerate backed rivals, who could match the service offered by Company Q while offering discounts of up to 15 per cent. At first the board dismissed this as a market entry strategy that would soon disappear, but in fact the new rivals became more price aggressive as they became further established, and also began operating to

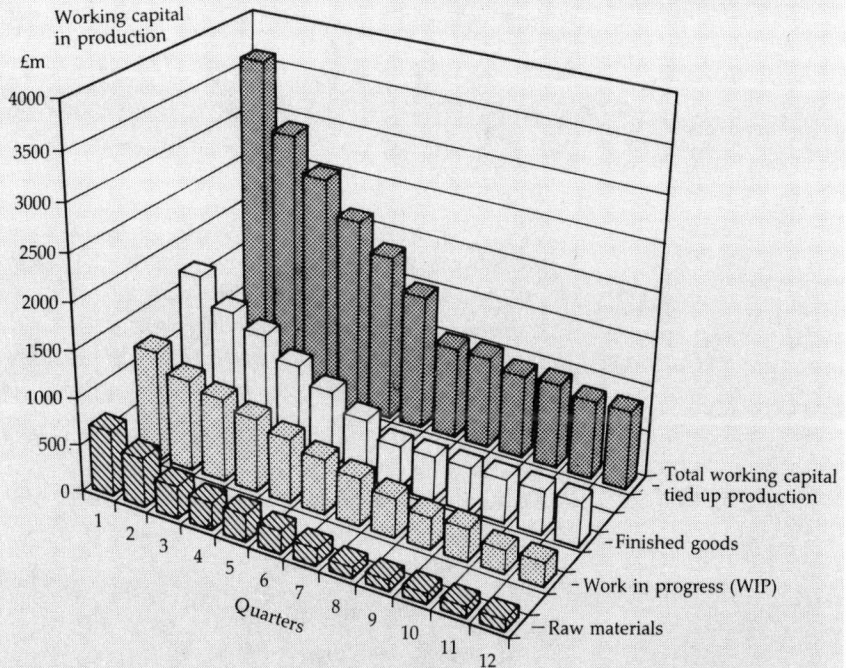

Figure 10.3 Changing working capital needs for Company Q

shorter delivery times with higher quality specifications. The board dismissed this as suicidal, but the rivals continued to gain market share and report growing profits. The board finally realized that urgent action was necessary when formally loyal customers began to ask Company Q to match the performance of these rivals.

After a series of consultations within the organization, and a report from a manufacturing consultant, the board produced the following objectives, to be achieved within three years.

1 Working capital to be reduced by 70 per cent;
2 Manufacturing lead times to be reduced from eight days to one day;
3 Floor space to be reduced by 30 per cent;
4 Indirect functions to be reduced by 50 per cent.

Looking at Figure 10.3, the actual cost savings can be seen. In Quarter 1, days of inventory for total capital were 73 days – based on a revenue of £17.5 million per annum. Management decreed that to stay in business the firm needed to overtake the competitor's lead – some 36.5 days. Production management were given the task of preparing and implementing a plan over a three-year period to do this. They set up project teams and employed all the other 'management by objectives' approaches discussed in this book. After considerable investment in machinery and many changes in working practices, management hoped to achieve, by Quarter 12, a total capital in production of £0.75 million. Based on the same turnover of £17.5 million, this was to yield results of:

17.5/0.75 million = 23 turns
per annum or 365/23
= 16 days.

Achieving the objectives

This was done by focusing on the constituent elements in which capital employed was tied up; notice that the reductions were by no means linear, or proportional to each other. For example, the best performance in reduction was forecast for raw materials, where the procurement team expected to be very effective in implementing JIT supply partnerships: The effect of this is as follows:

Quarter 1: Days of Raw Materials – which is based on annual materials consumed assumed at £8 million (a realistic comparison for raw materials, as buying effectiveness is measured to ensure minimum time tied up on the shopfloor) a comparison against revenue would be misleading.

Inventory turns for raw materials:

£8 million/£0.65 m = 12.3 times per annum

Expressed in days: 365/12.3 = 29½ days.

Quarter 12: Procurement's objective was to get this down to £8m/£0.1m (from chart) = 80 times per annum = 365/80 = 4½ days.

In practice the bulk of materials tended to arrive on the same day, but there was inevitable stockpiling of special materials required for say two to three weeks which put the average up.

Work in progress and finished goods were measured in terms of cost of sales, as they both contain an element of direct labour in them. The cost of sales was £15 million per annum.

WIP

Quarter 1: The ratio of WIP to cost of sales was 15/1.1 = 13.6 times per annum or 27 days – in other words nearly one month, which usually suggests a lot of 'dead time' between operations.

Quarter 12: The target capital in WIP was set at £0.4 million, a reduction of nearly ⅔, while still servicing all the variety demanded by consumers. In terms of the production challenge this is an increase in annual turns for WIP to 15/0.4 = 37.5, or under just under ten days.

This had two important hidden advantages:

1 The same floor space could cater for possibly three times the present capacity, were market demand to increase – so there was now in-built reserve capacity to handle 'peak loads' without delay.
2 The confidence levels that this built up with the marketing department and the company's customers boosted confidence in delivery capabilities without the need for carrying buffer finished goods – which often is a customer stipulation borne out of a need to be assured of continuous supply capability.

Finished goods

Quarter 1: Again, the ratio in Quarter 1 is 15/1.6 = 9 times per annum or a staggering 40 days. To bring this down to £0.5 million by Quarter 12 shows the sea change which JIT can cause to the company's marketing thrust – while saving capital!

Quarter 12: The ratio had fallen to 15/0.5 = 30 times per annum or 12 days – just under two weeks. There are companies now worldwide who are trying to bring this ratio down to two days – particularly when supplying bulk components. But, the effort to get under the two-week target sometimes yields diminishing returns, so that the effort and expense may well be better served elsewhere in the company – for instance by extra training.

These results were achieved through the introduction of FMS and JIT. JIT was applied following the use of Pareto rule; meaning that the most widely used components were organized in this way first. Most of the savings shown in Figure 10.3 were generated by control of less than one third of the components (measured by volume). Control of the remaining items was monitored more carefully than before. The space made available by closure of several storerooms was used to expand capacity and reduce bottlenecks.

Shorter manufacturing lead times were achieved by computer integrated manufacture (CIM), in which CAD/CAM systems were integrated with Computer Aided Design and Drafting (CADD) to enable faster design and manufacturing cycles. These changes, together with cell management in FMS have enabled the firm partly to close the production control function, thus reducing indirect production costs.

Quality improvements have been made by the introduction of robotics and Statistical Process Control (SPC) whereby output is inspected automatically at the point of manufacture, and instantaneous corrections are made to machine settings where deviations and defects are identified. These improvements have virtually eliminated the inspection of goods produced on site. New arrangements with suppliers have removed the need to inspect much of the incoming components and materials, further reducing indirect production costs.

These changes have not been achieved without considerable stress within the organization, with many operatives resisting the changes in working practices and conditions. In fact, at several points the whole process was threatened by high turnover of skilled workers at shopfloor level, and a refusal by some managers to delegate real power to cell operatives. Furthermore, many suppliers refused to change their standard conditions to accommodate Company Q's wish to buy small quantities at frequent intervals with short notice. In several cases, no local alternative existed and the further introduction of JIT has been somewhat restricted by this.

The fact that the process took three years was due largely to the difficulties in changing traditional attitudes and practices, rather than the technologies or systems involved. The issue of change management, and how these problems may be eased somewhat, is taken up in the concluding chapter.

In many respects, Company Q looks more secure now than it did in the mid-1980s. In fact, the company is trying to explore the many new marketing opportunities which have become available following the changes indicated. But the sheer size of the capital investment needed to generate these changes has caused a different set of problems. A small share issue now leaves the company vulnerable to takeover; indeed, the company is certain that at least one of its conglomerate rivals has abandoned the idea of breaking Company Q in the marketplace, and has switched its attention to takeover activity. Moreover, the heavy borrowing required has left the company highly geared. The new marketing opportunities must be taken up to ensure that sufficient profits are made to cover interest payments; but, at the moment, there is not sufficient capital available to finance them.

**Leap-frog
technology**

A company using this strategy will develop products and production systems based around leading edge technology. The idea is to be first, or one of the first, into creating a new market, initiating a whole new product life cycle, and gaining all the strategic advantages that may be won by being an early market leader (see Chapter 3).

It is often discoveries of new materials that finally bring ideas to life that previously have only been human dreams. Although the idea of 3D photography is a reality today, manufacturing methods, materials and processing technology are too complex and unaffordable to the mass market. But today's challenges become the next generation's standards, and when 3D photography does become a mass market product following further breakthroughs in process innovation, 2D photography is likely to disappear, just as mass market video camcorders have rendered the 8mm cine camera almost extinct in all but the specialist movie production industry.

Inventions by the research and development function abound in the search of new materials, be they liquid chemicals for 'cold washes', miracle easy-to-use car waxes, or new heat-resistant materials such as ceramics currently being tested as an option to the metals used in car engines. The patent protection afforded by the discovery and development of materials which outperform existing materials by an order of difference can make a traditional market obsolete almost overnight. For example, the transistor replaced the radio valve and revolutionized radios, giving them a new lease of life, and untold of market expansion as a result of the portable transistor radio, something just not conceivable with heavy but frail valves. Other well known examples abound, such as:

- Calculators replacing slide rules;
- Electronic ignition versus mechanical;
- Electronic eyes replacing mechanical metrology methods e.g. electronic scales replacing mechanical ones in every shop, and also being cheaper;
- Laser holography is still in its infancy, but already laser beams make cable TV a reality in combination with optic fibres since many TV channels, telephone and other signals can be simultaneously transmitted down the same beam, using the optic fibre as a transmitting medium.

These are examples of material and/or product inventions which are so far ahead of their rivals that they very quickly destroy the remaining life cycle of the traditional substitute.

The design engineer is at the forefront of converting innovations and good ideas into practical realities, targeted incisively for maximum competitive edge. Indeed, many innovative companies, such as those set by Sir Clive Sinclair, regard innovation rather than marketing as their distinctive competence, and follow a strategy sometimes referred to as *Me First*, where the aim is to be first to the market with a new product, and skim as quickly as possible in order to recoup development costs and finance the R&D for the next product before other firms enter the market. This issue was examined

Figure 10.4
Simultaneous
engineering

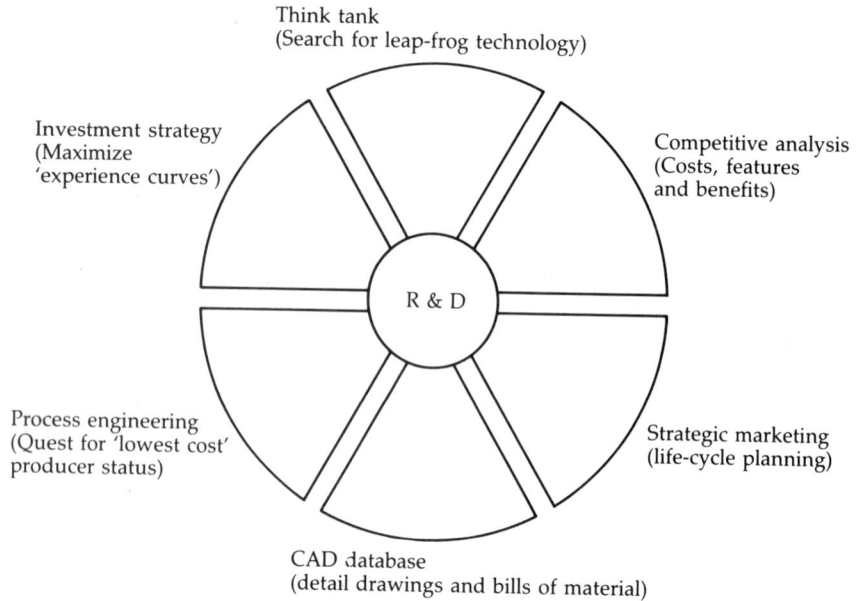

Think tank
(Search for leap-frog technology)

Investment strategy
(Maximize
'experience curves')

Competitive analysis
(Costs, features
and benefits)

R & D

Process engineering
(Quest for 'lowest cost'
producer status)

Strategic marketing
(life-cycle planning)

CAD database
(detail drawings and bills of material)

in Chapter 3; suffice to say that *Me First* companies can enjoy a stream of spectacular successes, but may fail catastrophically at the first major problem. The long-term development of a firm requires a thorough understanding of, and ability to work at, the interfaces of both the marketing and the manufacturing functions. As business moves towards the year 2000, and markets become ever more volatile, firms must develop an ability to innovate continuously as a necessary survival tactic, let alone for growth. This puts the engineer in prime pivotal position, although the need for rounded business management skills, such as finance and communication, is also crucial.

This diagram summarizes the key activities and issues at the forefront of the design process, and planning for the future.

Me Too

This phenomenon usually occurs once the pioneering of a new product or concept is finished and the market is in the rapid growth stage. The idea is to match closely the product offering of a near rival, or the market leader. It may be found in two distinct forms. First, it can be used as a market entry technique, where the best efforts and successes of the leapfrog entrants are duplicated, but the blind alleys and mistakes are avoided. In this sense it is a relatively low risk entry technique. However, it is not likely to work, on its own, in mature markets. IBM, for example, have not been at the leading edge of information technology hardware for some time, relying on setting an industry standard and incorporating tried and tested design features. As the market became ever more competitive, IBM began to struggle.

Me Too may be used to manage existing products in late maturity or end game, where successful innovations instigated by competitors are copied.

It would be most suited to products and services which the firm felt obliged to market to maintain a viable product range, but were not expected to gain market share or greater profits.

Me Too strategies are widespread in service sectors, and are not uncommon in erstwhile luxury markets such as high performance cars – namely Porsches with Japanese rivals such as the Toyota MR2 but at half the price (see Case Study 10.3).

Other examples include copying the hugely successful 'cornflakes' idea, originally brought out by Kelloggs, but now branded by many other companies; promoting the retaliation advertising slogan 'There is only one authentic cornflake'. The soft drinks and pharmaceutical industries are also fertile ground for this strategy.

On the technical side, patent protection is designed to encourage innovators to protect legally new concepts without fear of early plagiarizing. Once a patent period comes to an end however, copying of concepts is widespread, and even fatal to the future of the original innovators – who, through complacency, timidity or deliberate strategy may have under-invested in manufacturing processes so that other competitors waiting in the wings benefit from aggressive investment in updated processes. Examples include pharmaceutical, mechanized farming implements, industrial fasteners (for applications ranging from construction to aircraft manufacture) and many sophisticated component industries providing technology for white and brown goods consumer markets.

Me Too Plus

Similar to above, but in this case the copier will also add customer benefits to enhance well-established products, revitalizing them in, for instance, early saturation phases by boosting them into another growth phase. The most famous was the penetration of the UK and US motor-cycle industry by the Japanese – the Me Too Plus genius was in engine technology with high revolution two-stroke technology and lightweight bikes in the 150 to 250 cc capacity, creating a mass teenager market at the expense of larger four-stroke low-revolution engines whose main design features had often been around since before the Second World War.

These are examples of existing competitors being forced to innovate and invest in new processes in order to stay in the running; often a forlorn hope, either because the market does not forgive the originators for overcharging them for years, or because the originator has become bureaucratic and risk averse rather than innovative.

Lowest cost producer

This strategy relies heavily on manufacturing process innovation and ingenuity, so that extensive automation, usually involving robotics and flexible manufacturing systems, is applied to global mass markets. Cases in point abound, ranging from hi-tech components such as microchips, small DC motors for the car, white goods and office machinery industries to capital goods producers including shipbuilders and power generation plants. In all these applications, the Pacific Rim countries have shown a willingness to invest extensively in manufacturing processes with the long-term aim of global domination in targeted industry sectors. They are

Figure 10.5 Cost structures of two competitors. Which one will survive shakeout?

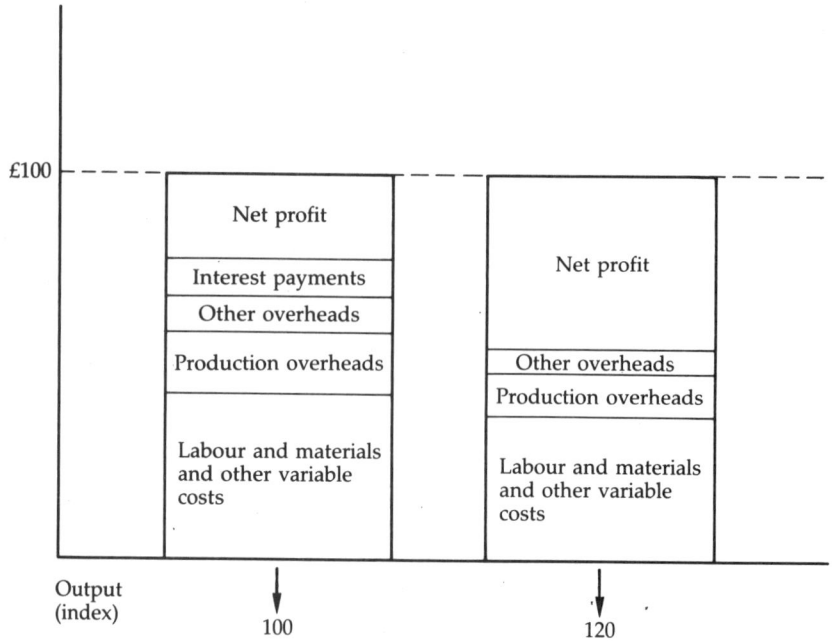

not alone however: Italy, Germany and the USA also have a reputation for lowest cost production in certain industry sectors, for example: air-conditioning systems in the USA; washing machines in Italy; and small DC motors, where the Germans rival the Japanese.

Traditionally, cost leadership has been regarded as a strategy which pays off in the shakeout and mature phases of the product life cycle, as there are clear advantages to the producer who can present the buyer with a more attractive product offering despite falling profit margins. Figure 10.5 shows two competitors with different cost structures; the implications for shakeout are readily apparent. However, a low-cost strategy can also open new markets. The successful Amstrad PCW wordprocessor range brought information technology into the price range of many new market segments. The low costs were based on a combination of modern production methods, cheap hardware, an operating system regarded as obsolete and ingenious software writers.

Case Study 10.3: Mercedes-Benz

Compact car with designs on success
Longer by a matchstick's length than its executive predecessor, and replete with two dozen new features as standard, the marketing motto for the Mercedes-Benz compact C-class is that it offers customers more car for their money.

According to Mr Klaus-Dieter Vöhringer, director in charge of passenger car production, buyers will get up to 20 per cent more value at prices basically unchanged from those of the out-going 190 series. Technically and cosmetically refined, the C-class is the fruit of manufacturing and management innovations with which Mercedes hopes to transform its prospects in the global quality car market.

It is the first practical example of the group's new pricing policy. The range embodies a principle new to Mercedes which states that before any work starts a new product will be priced according to what the market will bear and what the company considers an acceptable profit. Then each component and manufacturing process will be costed to ensure the final product is delivered at the target price.

Under the old system of building the car, adding up the costs and then fixing a price, the C-class would have been between 15 per cent and 20 per cent dearer than the 10-year-old outgoing 190 series, Mr Vöhringer said.

Explaining the practical workings of the new system, he explained that project groups for each component and construction process were instructed without exception to increase productivity by between 15 and 25 per cent. and they had to reach their targets in record time.

One result was that development time on the new models was cut to 40 months, about a third less than usual. But the most important effect, according to Mr Vöhringer, has been to reduce the company's cost disadvantages *vis-à-vis* Japanese competitors in this class from 35 per cent to only 15 per cent.

The time taken to build a car has been cut from forty-five man-hours to around thirty-five. Collaboration with outside suppliers has resulted in Mercedes making less of its own parts and also speeding up assembly. Complete electrical wiring harnesses, for example, are now supplied from outside on a just-in-time basis, ready to be inserted in one process into the bodies. Door interiors are also shipped complete for one-step installation. Where Mercedes used to produce 48 per cent of parts, it now makes around 42 per cent.

Dashboards and controls, built in a separate assembly shop, are also installed in one action instead of being bolted and screwed together piecemeal on an assembly line comprising rolling platforms in place of a conveyor system. Body assembly is robotized; side panels come in one piece rather than five parts.

Quality controllers have been moved from the end of the production line, where jams built up. They now work as members of each assembly team.

Group work has been introduced at every level of the Mercedes product cycle and six management layers have been thinned to four. According to Mr Vöhringer, this improved efficiency and collaboration with suppliers have each contributed 50 per cent of cost efficiencies.

The main task is to launch a new and still relatively costly car into crowded and depressed international markets. In Germany, which routinely absorbs almost half Mercedes' compact class output total industry sales are expected to fall 20 per cent this year. Yet the company is committed to producing 100,000 C-class cars and will step up production to 200,000 in 1994 compared with a recent annual average output for the old 190 series. Some executives even hope production can be cranked up to 250,000 in 1995, even though most forecasters reckon European car markets will still be weak.

In the meantime, Mercedes has to tackle the introduction of similar changes throughout the rest of its range. Three quarters of the passenger car output and most of its profits come from the middle-range executive classes and the luxury S-class. But it has ground to make up. Japanese competitors have a 15 per cent cost advantage over Mercedes' new compact model and are not renowned for standing still

Financial Times

Factoring

In Chapter 3 we argued that a firm creates value by identifying, making and communicating value. However, there is no requirement for all of the value to be made by one company. Value is created more efficiently if parts of the value creation process are handed over to specialists. In many senses, this is a non-manufacturing strategy since it involves using other companies to undertake part, or indeed all, of the manufacturing process. Providing the factor's margin is less than the cost saving involved, factoring should be seriously considered. Experience has shown that pre- and after-sales service, delivery times and customization services can often be factored cheaper and better than using 'in house' sources.

If a firm's distinctive competence lies in its distribution and marketing skills, and its reputation as a brand leader, there is no reason why it cannot enter, and dominate, a market with products made by a manufacturing partner. Indeed, nearly all 'own brand' products on the supermarket shelves are marketed in this way.

The tactic can be used to advantage in any stage of the product life cycle, but in competitive markets is almost essential. At the present time, we see many global companies trying to divest expensive in-house functions in favour of the marketplace.

Combined strategies

It is possible to combine strategies in some instances within a product group. A famous example of this was the battle on the world stage between Caterpillar and Komatsu in heavy duty earth-moving machinery for highway construction. Komatsu is now a dominant player in most earth-moving sectors through aggressive attention to developing lasting competitive advantage by delivering more economic benefits to the customer, by modelling themselves on Caterpillar as the benchmark of excellence that they needed to beat. This involved a combination of Me Too Plus *and* 'Lowest cost production' – the latter also involving innovative distribution

Figure 10.6 Changing structure of costs over the product life cycle

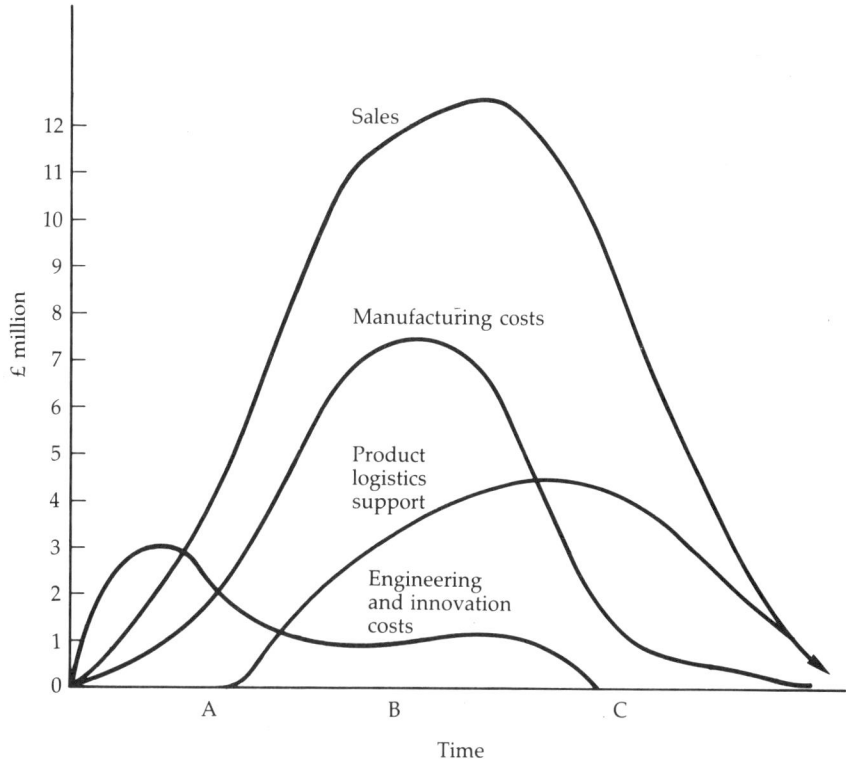

methods into the heart of Africa at unsurpassed low costs of distribution and service back-up. Caterpillar, with a reputation for premium service, could not stoop to meet its rival and were stuck with the 'Rolls-Royce' quality image for service back-up which they could not dare compromise – despite the fact that Komatsu soon equalled proficiency in service back-up, but allegedly at *half the cost.*

Strategic fit

Strategic fit is concerned with ensuring that the manufacturing and marketing strategies adopted by an organization are consistent with the resources and capabilities available to it. Discussion here will be based around the product life cycle.

We have suggested reasons why some manufacturing strategies may be more successful than others at various stages in the product life cycle. In some instances, it may be possible to change strategy as the market develops, but, as shown in Figure 10.6, the profile of principal costs will generally differ at different points in the product life cycle. Success in every stage of the product life cycle would require that the firm has a competitive advantage (in terms of cost and quality) in R&D, design, engineering and distribution at the innovative stages; manufacturing during middle to late stages and product logistics support at the later stages. But it is much more likely that firms will have greater advantages in some phases of the life cycle rather than others, depending on the skills and technologies available to the organization. Operating in a market which requires strengths

Think tank

*With managing, finance and
product directors:*
Competitive edge by creative design;
Weekly monitoring of new cost targets
against current bills of material costs.
Business planning with 40% margin.

*Communicate new business strategy
to whole company*

Investment strategy
with manufacturing and
finance directors:

Analyse investment costs of
production machinery and systems
as well as retraining needs and costs.

Ensure new suppliers meet
Quality standards and delivery
criteria (*Supplier audits*).

Prepare fabrication manipulators
for optimum costs from sub contractors.

Process engineering
Prototype buyer (an engineer)
analysing bought – in components;
and *setting* volumes for price
negotiations.

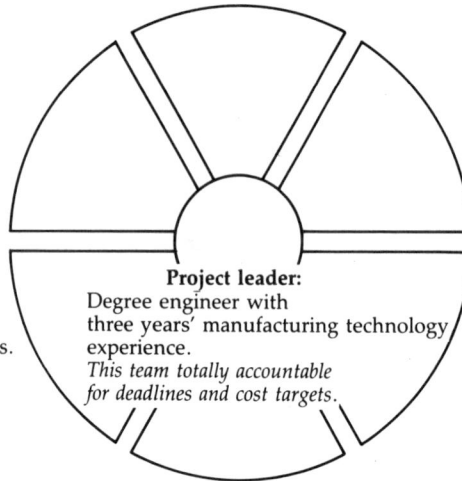

Competitive analysis
Reduce cost of new design
by 30% c/w existing models;

Design options add features to
obtain 10% price premium;

Analyse competitors;
machines for leapfrog
technology opportunities.

Strategic marketing
With marketing manager:
New literature and price lists;
Launch programme;
Exhibitions;
Volume forecasts for
manufacture;
Arrange secret prototype
testing with European dealers.

Project leader:
Degree engineer with
three years' manufacturing technology
experience.
*This team totally accountable
for deadlines and cost targets.*

CAD Database

Senior designers lay out models;
5 models tackled together for
"commonality"

Young CAD Engineers recruited for detail –
designing of whole range simultaneously.
– isolated to focus on development only.

Figure 10.7 R & D and product development, roles in change management: part of Case Study 12.2

unavailable to a firm will undermine competitive advantage. Company Q, discussed earlier in this chapter, may have won a manufacturing cost advantage through its investment activity. Consequently, the firm should be looking to develop this advantage by going for cost leadership through penetration strategies. This should influence the markets it tries to compete in, and the type of product it makes in those markets. A firm is said to have strategic fit if it is operating a strategy suitable for the stage of the product life cycle *and* consistent with its strengths and weaknesses. Since few firms have the capability to develop competitive advantage in all areas, most will have to pick the right moment to enter the right market with a marketing and manufacturing strategy which is appropriate for their financial, human and technological resources. As the market develops, firms should continually review their strategies, (including what to make and what to factor, and when to divest) whenever their strategic fit begins to break down.

As we have pointed out elsewhere, this mix of strategic and operational expertise, and multi-disciplinary approach to problem solving and implementation can only come about through a commitment to teamwork and change management. By way of example, Figure 10.7 is part of Case study 12.2 in Chapter 12, and shows interfaces between R&D and the rest of the organization necessary to secure chance of successful product development.

The difficulties of managing engineering interfaces was examined in the first chapter of this book. In the last chapter we explore further difficulties in managing change and the role of the engineer in leading project management.

Chapter summary

In this chapter we examined the means by which completive advantages can be created within the manufacturing function, first by controlling costs and second by becoming an integral part of the long-term strategy based around the product life cycle.

Further reading

Much of the cost cutting elements of this chapter are discussed in the management accounting references given at the end of Chapter 5.

Tutorial exercises

10.1 A casual observer might look at the electric motors case study on page 233 and conclude that competitive advantage is derived mostly from large scale production. Why is this misleading?

10.2 What steps could Company Q have taken to alert themselves to the problems earlier? Would it have made any difference? What should Company Q do now?

10.3 Figure 10.5 shows two competing companies in a market which is just becoming more price competitive. Use break-even analysis (see Chapter 8) to plot the existing condition of the two companies. How much profit would Company B make if it dropped the price to reduce Company A's margin to zero? What options would be open to Company A in such a case? Would you advise Company B to make this move?

10.4 Figure 10.6 shows industry average cost patterns and average sales figures for a particular product over the length of the product life cycle. Estimate and plot the profit made by a consistently average firm at points A, B and C. Contrast this with a competitor operating with a 25 per cent advantage in engineering costs, but a 20 per cent disadvantage in logistics support and average manufacturing costs; and a third firm with a 20 per cent advantage in manufacturing but average costs in other respects. What manufacturing strategies would be best advised in each case, assuming that the market is price sensitive in the mature phase? Make clear any assumptions you may need to make about the nature of the costs concerned.

10.5 Consider Case Study 10.3. What generic and manufacturing strategies have recently been adopted by Mercedes-Benz? What will Toyota do? What eventual outcomes do you predict?

10.6 (a) Suppose Company Q were to opt for an innovation based strategy, based on short production runs and a skimming strategy. What problems might occur?
(b) Suppose Company Q's investment effort had not created a competitive advantage in lower manufacturing costs, but had simply matched the cost structure of the conglomerate rival. What options and problems suggest themselves?

11 Long range financial planning

In the last chapter, we looked at the decisions which affect the growth and development of the business. In the final chapter, we examine the organizational changes that may be required by the future plans and missions of an organization. In this chapter, we examine the formation of those plans from a financial perspective, in particular, how marketing, engineering and accounting data is used in investment appraisal techniques. First in Figure 11.1, we examine investment appraisal techniques, and, second, consider how to estimate the cost of capital for a project, and the impact of inflation.

Engineers and managers in today's operating environment need to recognize the realities of resource constraints and the financial consequences of engineering activity. This does not mean that the role of an engineer is subordinated to financial needs – merely a recognition that when selling ideas and projects to non-engineers, there needs to be a common understanding which is stated in financial terms. An example of common understanding has already been discussed. In Chapter 6 we examined ROCE, a key ratio used for improving overall financial productivity of the organization. Similarly, the effect of new engineering activity on 'the bottom line', more correctly called net profit, must be assessed before the project is undertaken and the cash flow implications of capital investment should be estimated over the life of the project.

The engineer should have an understanding of how senior managers assess project and investment decisions so that engineering efforts are not

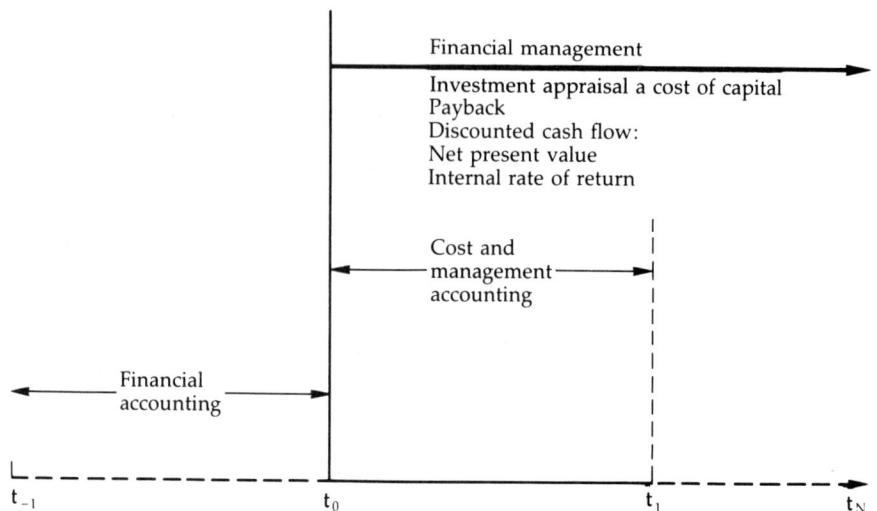

Financial management

Investment appraisal a cost of capital
Payback
Discounted cash flow:
Net present value
Internal rate of return

Cost and management accounting

Financial accounting

t_{-1} t_0 t_1 t_N

Figure 11.1 Phase 3: investment appraisal in context

wasted, and good projects can win support with an adequate budget. This chapter will outline the basis of this common understanding by exploring the final component of the model illustrated in Figure 5.1. We will examine ways organizations can seek to appraise possible projects and select only those which will fulfil objectives and add financial security.

The nature of capital investment decisions

In Chapter 9, budgeting in organizations was described. The budgeting model indicated that a number of subsidiary budgets were needed before the overall budgeted profit and loss, balance sheets and cashflow statements could be formulated for the total organization. One subsidiary budget identified was the capital expenditure budget. This budget is critical if long-term competitiveness of the organization is to be financed.

All major purchases of capital equipment and plant will arise from this budget, as will equipment that improves efficiency. The details surrounding capital expenditure are dependent on the nature of the firm and industry, for example, CAD/CAM technology in a manufacturing firm, laptop computers for all sales engineers, replacement of the central mainframe computer; installation of safety equipment as recommended by the Health and Safety Executive and pollution control systems.

The major features of capital expenditure budgets are:

1 They often involve relatively large items of cash expenditure that have a major impact on cash-flow statements;
2 The items included in the budget have been formally agreed in advance by senior managers and directors with authorization to commit large items of expenditure;
3 The capital expenditure reflects the overall longer-term direction of the organization;
4 The nature of capital expenditure decisions involve major repair programmes, replacement, new investment in buildings, plant and machinery, and information technology expenditure.

However, capital expenditure needs to be financially justifiable. Committing large cash resource in the next financial period could have marked effects on the liquidity of the firm. But, the dilemma all firms face is that by not committing funds to R&D, new product development and launch, and updating capital equipment (see Chapters 3 and 8) there is a major danger that competitive edge will be lost, and the ability to survive in existing or new markets will be impaired.

Capital expenditure is thus synonymous with investment. Firms involved in producing new technology products or systems are often committing large funds now to enjoy profitability, or at least survival, in future periods of time.

Engineering investments

Engineering businesses, by their very nature, do not operate in secure risk-free environments. The capital raised for an engineering project has to provide returns to the capital providers (investors) commensurate with their overall attitude to risk.

For example, suppose a capital provider had a choice of investing in one of two projects: *either* earning 10 per cent in a savings institution, such as

Figure 11.2 Stages of
project development

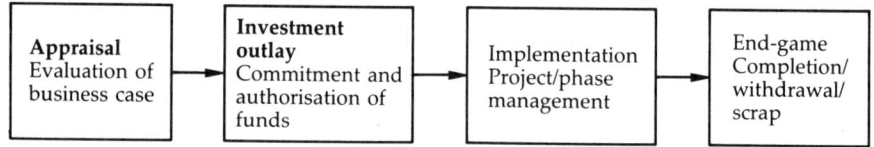

| Appraisal Evaluation of business case | → | **Investment outlay** Commitment and authorisation of funds | → | Implementation Project/phase management | → | End-game Completion/ withdrawal/ scrap |

a building society, with very little risk for the next few years; *or* committing the same funds to a civil engineering project involving tunnelling into fifteen kilometres of a granite mountain to provide a toll road, which might give a return of 10 per cent (over the same period as the saving institution).

There is unlikely to be indifference between the relative risk-free investment of the savings institution and the high risk investment of the tunnel project, if both are offering the same return. However, if the latter were offering a return of 30 per cent the decision might not be so clear cut. It would depend on the investor's attitude to the higher risk involved. The 30 per cent may be sufficient as an incentive to induce the potential investor to commit the funds to the project as opposed to the relatively risk-free investment in the building society which provides the lower return. Of course, other investors would find the risk involved in the project to be too high, and opt for the safer option and lower return.

In any engineering project there are a number of phases from conception through to implementation and eventual project completion. These phases are listed below:

- Appraisal
- Investment Outlay
- Implementation
- End Game.

The nature of the industry would indicate the time-scales upon which the returns on investment begin to accrue and for how long. Major civil engineering projects would have much longer cycles than, say, computer hardware developments. The product life cycle referred to in Chapter 3 would suggest that the benefits of any single investment project in consumer goods, and in other engineering markets like armaments industries are finite. Some civil engineering projects, or major bespoke projects may have a much longer lifespan, although the risks may be even higher and the period before any benefit is recouped may be even longer than with consumer goods. It is worth reminding the reader of the important links between the product life cycle and cost implications discussed in Chapter 10. In this chapter, we are mostly concerned with project appraisal.

Project appraisal

In an innovative company, a large number of exciting new ideas will be generated simultaneously. Appraisal techniques are used by senior management to select the best projects for development and implementation. There are a number of techniques that can be employed to appraise the necessary financial commitment and the future financial benefits that will flow as a consequence. The techniques all have particular strengths and weaknesses.

They can be made simple or complex depending on the level of sophistication that is desired. Mathematical complexity becomes less of a problem with the use of computer applications using commercial modelling simulations.

The techniques can be broadly split up into two groups, firstly, those based on cashflow projections, and secondly, those based on accounting profit estimation. The techniques based on cashflow will be examined first in the following order: payback; discounted cashflows using both Internal Rate of Return (IRR) and Net Present Value (NPV). Techniques based on accounting profit are examined later in the chapter.

Techniques based on cash flow

Payback method

The appropriate data to use for any payback calculation is projected cashflows. If suppose a current engineering project requires an investment of £500,000 in t_0. Then, if the project is to be worthwhile, the financial benefits that arise as a consequence should be greater than £500,000 over time. If the positive cash flows are £150,000 in t_1, £200,000 in t_2, and £300,000 in t_3, simple addition tells us that:

£150,000 + £200,000 + £300,000 = £650,000

Because £650,000 > £500,000 the investment appears to be justified over the project life.

But if the condition was made (by the engineering director) that for the project to be viable the investment must payback within two years, then:

£150,000 + £200,000 = £350,000

Thus, the proposed investment fails on financial criteria because £350,000 < £500,000.

The exact payback period (or date), is calculated over the following four stages:

1 Sum the annual cash flows
Cumulating the cashflows from the initial investment and for each successive time period:

Time (t)	cashflow £	cumulative £
t_0	(500,000)	(500,000)
t_1	150,000	(350,000)
t_2	200,000	(150,000)
t_3	300,000	150,000

2 Identify the year in which cumulative cashflow becomes positive.
The payback first becomes positive between the second and third year.

3 Estimate the exact month in which cumulative cashflow is zero.
Assuming an even rate of cashflow over the year, the payback month is:

$$\frac{\text{last negative balance in the cumulative total}}{\text{cash inflow in the next period}} \times 12$$

$$\frac{(£150,000)}{£300,000} \times 12 = 6 \text{ months}$$

4 Add year (stage 2) and month (stage 3).
The payback period is two years and six months.

The technique is very simple. It is used widely by many businesses. It can be used in conjunction with other more sophisticated techniques. However, it does have severe shortcomings, including:

1 It is 'short-term' by nature. Any returns beyond the required payback period are ignored. Thus, as in many longer-term engineering projects, the early poor returns from an investment are likely to fall foul of the payback rule used by that particular business;
2 If payback were to be used in a longer-term context then it would be inferior to discounted cashflow techniques by ignoring the time value of money (see next section);
3 It may encourage short-term investment because the criteria is met but those future cashflows which have been ignored beyond the payback period may be negative or prove negative when discounted;
4 It ignores 'end-game' effects of discontinuing the investment, scrapping costs and salvage revenues, and any tax implications.

Nevertheless, the method is a popular 'rule-of-thumb' approach to investment decision-making when so many projects are often having to compete for a finite level of investment capital within the same business. Unfortunately, when the mainstream core business relies on engineering development and marketing then this type of measure could, (and some would argue, commonly does) stifle long-term engineering initiatives.

The time value of money

The payback technique assumes that £1 today is worth £1 in one year's time. The same assumption suggests that the £1 in one year's time is worth the same as the year after, and so on. In fact, conditions are unlikely to be true. Consider the case of a small investor offered the choice of £100 now, or £100 in a year's time. The investor will always choose to take the money now, for a variety of sound reasons. First, he or she may die before the year is up, in which case deferment would have no value at all. Similarly, the offerer may be unable to provide the £100 in a year's time. However, deferment would not suit our investor even if these situations do not occur, since inflation in the economy will erode the purchasing power of the £100 before it is received. But even if inflation were zero, the investor will still lose out by deferment, because the £100 could have been invested with little or no risk and would, therefore, have been worth more than the £100 offered after one year.

Generally speaking, money is worth more now than in future, because of risk, inflation and loss of investment opportunity. And the longer the time period under consideration, the greater the erosion in the value of money. Payback criteria ignore these factors, but more rigorous assessment criteria would have these built in. In the next section, we examine how this is done with the principles of compounding and discounting.

Suppose the small investor deposited the £100 in an investment account, with an interest rate return of 10 per cent. There will be a return of £10 in one year's time. The £100 deposited in t_0 will be worth £110 in one year, t_1. Thus, the reward for taking £100 now, rather than in 1 year, would be £10.

If it was decided to invest the £100 in one year's time for another year then, assuming the interest rate was to remain the same, the investment would be worth, in t_2:

£110 $\times$ (1 + i) = £121 where i = 10 per cent

Similarly, if invested in year 3, the return will be

10 per cent $\times$ £121 = £133.10

Hence, investing any principal sum, P, for n years will become

$P (1 + i)^n$

For example, if P = £100, i = 10 per cent, and n = 3 years, the sum would become £133.10. These are the principles of *compounding*.

So, to return to our investor, in order to make the deferred choice attractive, we would have to compound the original £100 up to £110, at least. Alternatively, we could continue to offer £100 in one years time, but offer less if the investor chooses to take the money now. We estimate this reduced sum by the process of *discounting*, which is the reverse of compounding. In order to make deferment attractive, we should have to offer the sum which would generate £100 in one years time, at the interest rate of 10 per cent.

The sum is given by the formula:

$$\frac{£100}{(1 + i)}$$

where i = 10 per cent, then

£100/1.1 = £90.90

Therefore a more realistic approach to encourage the investor to consider deferral would be to offer a choice of £90.90 now, or £100 in a year's time to make the options of equal value.

The question of when to discount is raised when investment decisions are being considered at the present time which may involve committing relatively large sums for the future life of a project.

The benefits of the investment can be estimated over the expected life of the investment. But those estimates would need to be compared with the initial investment being made in t_0. Discounting t_1 to t_0 will give the same result as compounding to t_0. So taking our earlier example of a three-year investment in a 10 per cent account:

The value now of £1 arising in 'n' years time, when the annual rate of discount is 'i' is given by:

$$\frac{1}{(1 + i)^n} \quad \text{or} \quad (1 + i)^{-n}$$

If the question was asked as to how much the £133.10 in three years time would be worth now (t_0) assuming a discount rate of 10 per cent then, using the expression above:

$$£133.10 \ (1 + 0.1)^{-3} = £100$$

In summary, by discounting, we can remove these time effects of money by discounting future investment income. In the example above, we have used the rate of interest to compensate for the investment income lost by deferral. However, we could include inflation and risk by increasing the discount rate.

It is this logic which is applied to Discounted Cash Flow (DCF) techniques of investment and project appraisal. The net present value (NPV) and internal rate of return (IRR) are two specific DCF techniques which will be discussed briefly.

Net present value (NPV)

If future discounted cash flows as a consequence of the investment are added together and are greater than the original investment (capital outlay), then the business will be better off by making the investment.

There are essentially five key inputs to the NPV model that need to be identified over a discrete time period:

1 the discount rate to be used for the project investment;
2 the initial outlay;
3 the future cash inflows arising from the investment, and;
4 the future cash outflows arising from the investment;
5 the time period of the project.

For calculation purposes there are a number of approaches that can be taken, including arithmetical progression and tabulation discussed here. The former method can be expressed thus:

$$\sum_{t=0}^{n} \frac{A_t}{(1+r)^t} = NPV$$

where
A_t is the project's net cash flow per period over the life, n, of the project (t takes on the value 0 to n);
r is the annual rate of discount. It is assumed in the expression that r will remain constant over the life of the project.

For demonstration and presentation purposes, a tabulated approach will be used because the data inputs to the model can be examined.

Sticky Sweets, and its competitors (see Chapter 3), have all seen sales literature concerning a specialist assembly line costing £150,000. The line is expected to last four years. The net cash inflows to the project over the four years are £45,000 per annum for the first three years, and £35,000 for the final year. (The net cash flows are positive each year because the savings in labour offset the cash costs of operating the machine.) The discount rate to be used is 16 per cent.

			(£000's)		
	$t0$	$t1$	$t2$	$t3$	$t4$
Investment outlay	(150)				
Net cash flow		45	45	45	35
Discount factor 16 per cent	1	0.86	0.74	0.64	0.55
Present value cash	(150)	38.7	33.3	28.8	19.2

NPV = (£30,000)

The net cash flow from the investment is £170,000. If we subtract the £150,000 of original investment, the result is a positive cash flow of £20,000. But this ignores the time effects of money discussed above. The present value of cash shows the future income discounted to its value now (or present value) by the same process as our example above. By summing the yearly present values we find that the project has a negative NPV of £30,000.

Decision: As the NPV is negative then the proposal should be rejected on financial criteria. Sticky Sweets will be worse off by making this particular investment.

Activity: Sourr Sweets is a foreign competitor of Sticky Sweets, who has also seen the literature relating to the new assembly line and the cash flows arising from the investment. Rework the above example using a discount rate of 4 per cent, then see tutorial exercise 11.3 at the end of the chapter.

Answer: At a discount rate of 4 per cent, the project generates a positive NPV of £4,400. For illustrative purposes, the answer is derived below using a progression.

£000's

$$\underset{t_0}{\frac{(150)}{1}} + \underset{t_1}{\frac{45}{(1.04)^1}} + \underset{t_2}{\frac{45}{(1.04)^2}} + \underset{t_3}{\frac{45}{(1.04)^3}} + \underset{t_4}{\frac{(45-10)}{(1.04)^4}} = £4.400$$

Tables with discount factors already calculated are readily obtained, and can obviate the need for long and repetitive calculations, as long as the lifetime of the project and the discount rate can be estimated.

The decision rule using the NPV technique in this case was to reject the proposal on financial criteria when the discount rate was 16 per cent. The investment outlay cannot be justified because the risks taken by the investors (capital providers), reflected through the discount rate, are not sufficiently rewarded.

But when the discount rate was lower at 4 per cent, the cash flow was positive. The lower discount rate might have been selected because the company could raise the investment capital more cheaply, or because its managers were less risk averse, or because its investors have lower requirements. Further, the competitor may be based overseas, where inflation and/or interest rates are much lower. Hence, recurring problems of high inflation, high interest rates and slow growth will adversely affect investment decisions in relation to overseas rivals.

The choice of discount rate is critical to whether an investment in an engineering project is made or not. Indeed, all the input variables require careful consideration when proposing and justifying projects. The matter of the discount will arise again in the next DCF technique to be considered, the internal rate of return (IRR).

Internal rate of return (IRR)

This method requires an understanding of the NPV technique. The NPV will express, in absolute '£' terms, a positive or negative figure which, in part, depends on the choice of discount rate. The practical use of the IRR is to determine the rate of return, in percentage terms, that will be required to ensure that the NPV of a project is equal to zero. It is this rate which is then compared with the rate for other projects with a similar risk profile. If the project is to be considered in isolation, 'Does the project under consideration yield a better return than a given "hurdle rate" given by the business which, in turn, should be a reflection of the requirements of its external investors?'

The IRR of a project is given by:

$$\sum_{t=0}^{n} \frac{A_t}{(1+r)^t} = 0$$

Where

A_t is the project's net cash flow per period over the life, n, of the project (t takes on the value 0 to n);

r is the project's approximate rate of return and can to be solved using interpolation.

In the NPV example, a 16 per cent discount rate yielded a negative NPV and the 4 per cent discount rate gave a positive rate of return. The point at which the NPV = 0 can be plotted on a graph:

At 4 per cent the NPV is £4,400; very near to the point where the NPV is zero. The exact discount rate will be where the graph intersects the horizontal axis and, if plotted to scale, a reading can be taken. If solving without the use of the graph then by using linear interpolation an approximation can be sought:

$$4\% + \frac{4.4}{4.4 - 30} \times (16\% - 4\%) = 5.5\% \text{ (approx.)}$$

Therefore, the IRR of this project is around 5½ per cent. If the hurdle rate was less than 5½ per cent, then the project is acceptable, but not if the hurdle rate is higher. This example is shown in Figure 11.3.

Techniques based on accounting return

All the techniques examined so far have been based on cash flow projection. However, it is important to recognize the distinctions between cash-flow and the kind of accounting profit statements described in Chapter 6.

The investment appraisal techniques discussed so far have been projected cash-flow data. There are valid reasons for doing this. For example, accounting profits are affected by issues of timing. The accruals concept is used to measure profit in that period which includes an assessment of the use of the resources even though they may not have been paid for in cash by the end of the accounting period. Similarly, depreciation – a non-cash cost – reduces the accounting profit in the periods for which it is charged over the life of the asset. If the asset to be depreciated was purchased with

Figure 11.3 NPV and discount rate

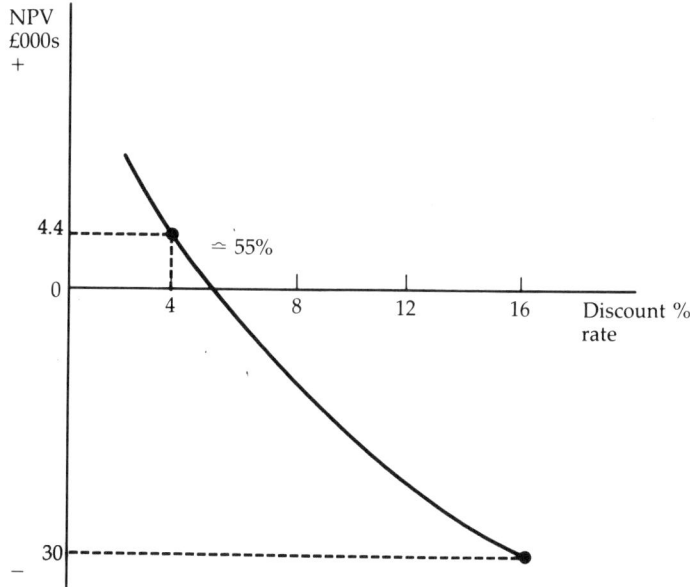

cash, then notional depreciation was charged, in cash-flow terms, when purchased. For these reasons, an investment criteria drawn from a profit figure based on accounting conventions could be misleading.

However, the Accounting Rate of Return (ARR) is an alternative method, based on accounting profits (and criticisms of it as a technique are based on this fact).

The accounting rate of return (ARR)

Unlike payback (and subsequent DCF techniques) this method uses accounting information for project appraisal – not cash-flows. Alternatively the ARR is used interchangeably with the ROCE, which has already been discussed in Chapter 6. The ROCE encountered to date was historical where the capital employed was extracted from the balance sheet and the operating profits from the PLA. For investment appraisal purposes, the accounting profits would have to be projected from a capital outlay (investment).

For example, Rally Engineering wished to evaluate an investment proposal of £100,000. It is anticipated that the project will generate cash flows of £40,000, £50,000, £30,000 and £20,000 in years t_1 to t_4 respectively. The investment will be depreciated using straight-line depreciation with no residual value of the project at the end of its life. There are a number of steps that need to be taken.

Step 1: Calculate the accounting profits each year and to determine the capital employed for each year:

	t_0	t_1	t_2	t_3	t_4
			£s		
Cash flow		40,000	50,000	30,000	20,000
Less: Depreciation		25,000	25,000	25,000	25,000
Profit		15,000	25,000	5,000	(5,000)
Capital employed	100,000	75,000	50,000	25,000	

Step 2: Calculate the average profit generated over the life of the project:

$$\frac{(£15,000 + £25,000 + £5,000 + (£5,000))}{4 \text{ years}} = £10,000 \text{ p.a.}$$

Step 3: The average profit could be expressed in one of two ways, either a percentage of the initial outlay or as a percentage of the average annual investment. Both are illustrated below.

(a) As a percentage of the initial outlay of £100,000

$$\frac{£10,000}{£100,000} = 10 \text{ per cent}$$

(b) As a percentage of the average investment over the life of the project. The average investment in the project is simply the mid-point investment level over the life of the project:

$$\frac{£100,000 + 0}{2} = £50,000$$

Therefore, the ARR in this case is $\dfrac{£10,000}{£50,000} = 20$ per cent

The ARR is widely criticized, particularly when compared to DCF, because of its tendency to use accounting profits. As in the example above, the depreciation policy and residual values will have an impact on the overall ARR – particularly if measured annually over the life of the project.

However, many organizations, and their managers, continue to use it, albeit in conjunction with other methods. The reasons are primarily concerned with management problems similar to those with payback: first, the measure is expressed in a simple percentage and this can easily be compared to the historical ROCE (rightly or wrongly); and secondly, the increasing tendency to reward engineers and managers by results is often tied to improving the ROCE on behalf of shareholders.

The cost of capital

So far, we have focused on generating cash or profits over and above the original cost of the investment. Now we examine in greater detail the effects of raising the finance necessary to fund the initial investment. First, we will briefly review balance sheet implications, and then look at capital financing for a given project. In particular, we will look at the implications of capital funding on the balance sheet.

A firm's balance sheet is unique. One of the reasons it is unique is that the capital composition of owners and lenders will differ. The availability of capital available to a business organization is a severe constraint in most cases. Individuals generally have only limited resources. Banks and capital markets have very well defined criteria to meet before they can be accessed for resources.

In Chapter 6, the N. Grineers example on page 110, the creation of a balance sheet was caused by external funds being made available to the business.

Figure 11.4 Structure
of funding

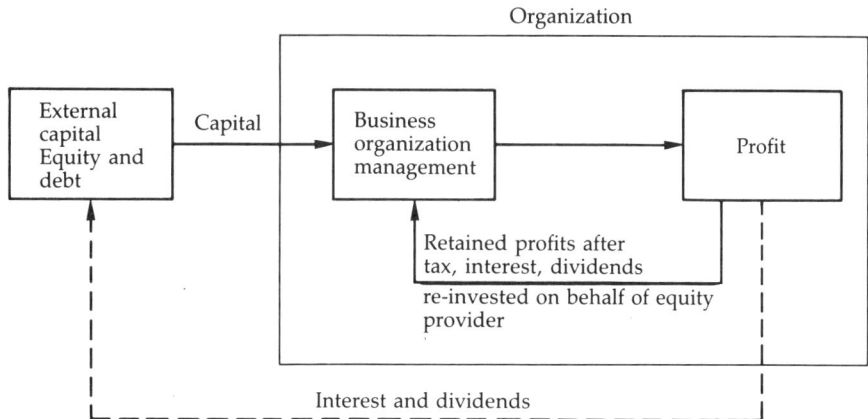

Funds were placed at Mr and Mrs N. Gineer's disposal to execute their roles
as engineers and provide a product or service. If it were their own capital
then, should the business fail, it is their own funds that are lost. Equally,
should the business prosper then they will be the main beneficiaries.

If the capital were not all their own, then there must be the offer of some
reward that will satisfy those who contributed some or all of the capital
(increasingly in the public sector, capital would be substituted by fund-
holding authorities or agencies to which the management of the organization are accountable).

The composition of any business's balance sheet will indicate the make-
up of capital providers, as examined in Chapter 5. There are broadly two
groups of capital provider:

1 debt, and;
2 equity (shareholders).

The debt-provider often advances capital for a fixed period of years in return
for a fixed interest payment annually. At the end of the period the capital
(or principal sum) is often refunded. The equity-provider often advances
capital for an indefinite period of years in return for a possible floating return
in the form of dividends. (The stock market in the UK provides a ready
market for 'plc' shareholders to sell their shareholding to the public without
the constraints imposed upon private company. The share price quoted is a
reflection of buying and selling activity for that company's shares.) Both
interest and dividends are paid out of profits. But interest must be paid out
of profits (and cash-flow) before any payment can be made to shareholders.
Hence, when profits are secure, there is little risk to the debt-providers – they
receive their interest payment regardless. If profits grow then the shareholders are the beneficiaries of the profit growth. But if profits fall, the debt-
holders will continue to expect their interest payment nonetheless. The
shareholder, however, may not receive any dividends for that year. The
shareholder takes the greater risk over time and will therefore, wish to be
rewarded at a premium over and above the debt-provider.

This allocation of risk and reward between owners and lenders is the
acknowledged basis upon which capital providers invest their funds in
companies. Generally, the more susceptible a company is to swings in

profit or loss, the greater the return in the form of dividends (and/or capital growth) expected by shareholders over time.

There is a substantial body of theory concerning whether or not there is an optimum mix of debt and equity in a company. If debt is cheaper than equity it would make sense for the company to employ as much debt as possible. Too much debt exposes the company to risk that may increase the shareholder's required returns (whether there is an optimum mix of debt and equity is beyond the scope of this book).

A key solvency ratio outlined in Chapter 6 was the gearing ratio, expressed as debt/equity, which indicates the potential exposure to risk that a business faces should there be a downturn in trading and profitability. The higher the gearing ratio, the more a company needs to to seek high profits in the short-term in order to pay interest. The case of Company Q in Chapter 10 is typical of a company which has become too highly geared.

At project level, the gearing issues are even more problematic, and several techniques are used to assist in capital planning and project appraisal.

The weighted average cost of capital

Discount rates can be set quite easily when capital is sourced from one supply, i.e. all debt or all equity etc. since the cost of the capital is relatively easy to identify. However, in most cases, capital will be drawn from several sources, and an average cost of capital could be estimated in order to set a discount rate to be used in the calculation of NPV.

One technique that can be employed to determine the discount rate for an NPV is the Weighted Average Cost of Capital (WACC). The reason for using WACC as the basis of the discount rate for investment proposals is that by financing a project yielding a positive NPV using combination of debt and equity, the total value of the firm will be increased. WACC is based on the amount of capital raised from each source, and the cost of each source.

Example of a WACC

An electrical engineering company, Hyland Electrical Contractors Ltd, is financed by equities, preference shares, debt and a mortgage on the premises.

It has 1,000,000 ordinary shares issued at par with a price of £2 each, paying a £0.40 dividend per share; 500,000 preference shares, par value and issue share price of £1, providing a dividend rate of 12 per cent annually. In addition there were £600,000 10 per cent debentures issued at par; and a mortgage of £300,000 bearing 14 per cent each year. The tax rate is 40 per cent.

Required:
Calculate the WACC for Hyland Ltd.

Solution:	(1) £ Amount	(2) % weight	(3) Annual Dividend Interest £	(4) Annual After-tax Cost £
Ordinary shares	2,000,000	59	400,000	400,000
Preference shares	500,000	15	60,000	60,000
Debentures	600,000	17	60,000	36,000
Mortgage	300,000	9	42,000	25,200
Total capital employed	3,400,000	100	562,000	521,200

1 Because the company's shares are not quoted, the information for the basis of WACC calculation will be taken from the balance sheet, using book values. If the company were listed then the UK market values of the debt and equity instruments would be known and the WACC could be calculated on the basis of market valuation.

2 The proportion of each type of capital instrument (long term debt and equity) is considered and weightings calculated.

3 The returns are calculated on each category with the information given.

4 Because loans are tax deductible for businesses in the UK, both debentures and mortgages are calculated to provide the after-tax cost by multiplying by $1 - t$, where t is the rate of tax.

5 For any NPV calculation, the weighted average cost of capital after tax, k_0, will be:

$$\frac{£521,200}{£3,400,000} \times 100 = 15.32 \text{ per cent}$$

6 Therefore, the company would use a discount rate of approximately 15 per cent in its project appraisal.

There have been many criticisms made of the WACC approach to determining a discount rate. These criticisms, and alternative approaches to the problems, are well covered in the first reference cited in 'Further reading' below.

Inflation and investment

It was stated earlier that the time value of money was somewhat broader than the notion of inflation. However, inflation can undermine the integrity of financial information and the decisions arising from it unless its existence is acknowledged at the planning stage.

In Chapter 5, the stability concept of financial accounting, assumed the purchasing power of the currency will remain constant year on year. Inflation doesn't invalidate the accounts, it simply means treating them with a little more circumspection than would be done otherwise. Of course, the accounts of a firm ten years ago compared with the current year will be of little relevance if there has been a history of inflation in the period.

In a DCF model, projected inflation can be built in by:

1 Adjusting the discount rate upwards (if assessed as rising) by the rate of inflation plus;

2 Adjust the cash flows using the projected rate of inflation and provide monetary values.

Chapter summary

Engineering projects frequently transcend the short term. Before engineering work is authorized, the financial implications are subject to increasing scrutiny. This chapter has identified and described the techniques most frequently used in justifying capital expenditure.

Further reading

The basic investment appraisal techniques are developed in *Accounting for Management Decisions*, by J. Arnold and T. Hope (1990), and in T. Lucey's very useful book, *Quantitative Techniques* (DPP, 1988). Appraisal, and finance generally, is linked to the strategic issues raised throughout the book in *Corporate Resource Allocation*, by C. Tompkins (Blackwell, 1991). The behaviour of management and organizations in dealing with uncertainty and risk is well covered in *Making Management Decisions*, by S. Cooke and N. Slack (Prentice Hall, 1991).

Tutorial exercises

11.1 Company Q (see previous chapter), wishes to diversify into volume production on a newly engineered safety device that could be sold worldwide with a patent. Market research has indicated that this is a very competitive market and the product is sensitive to price:

Price £/unit	year 1 (units)	year 2 (units)	year 3 (units)	year 4 (units)
100	40,000	50,000	60,000	50,000
80	80,000	90,000	90,000	70,000

Their findings are the basis of discussions between the marketing, engineering, finance and production directors whether to invest in the higher volume production or the lower volume production facilities.

Low volume production will involve a relatively low initial outlay of £2 million, but incurs higher variable costs of £80 per unit with lower fixed costs £200,000 for four years.

High volume production will require a higher initial outlay in advanced technology of £4 million, incur lower variable costs of £55 per unit and higher fixed costs of £600,000.

It is assumed that the realistic life of the project will last for four years. There will be no residual values for the machinery. A suitable discount rate for investment in this venture will be 16 per cent.

Required:
(a) Evaluate using NPV whether to invest in the high production machinery or the low production machinery.
(b) What other information might be considered before the decision to invest is made?
(A suggested solution is shown on page 264.)

Case study 11.1: Volkswagen & SMH

VW quits small car venture with Swiss group

By Kevin Done, Motor Industry Correspondent

Volkswagen, the German car maker, has been forced to abandon its pioneering small car development project with SMH, the Swiss watchmaker.

VW, under financial pressure with profits falling, is pulling out to avoid heavy investment in the project. SMH said VW was withdrawing because it had to cut costs.

The Swiss company, maker of the popular Swatch watch, said it would continue with the project and was considering offers and alternative partners.

VW joined forces two years ago with SMH to develop a small electric car for city use. It is continuing development work on its own city car project, called Chico, which may be powered by a hybrid petrol/electric power or by a two-cylinder engine under development. General Motors, the US car maker, said yesterday that sales of its Opel/Vauxhall cars in Europe reached a record 1.61m units last year, a 3.9 per cent rise from 1.55m in 1991. GM said its cars, which are sold under the Opel brand in continental Europe and Vauxhall in the UK captured 12.0 per cent of west European new car sales in 1992 compared with 11.6 per cent a year earlier.

Total GM group car sales in west Europe (including Saab, Lotus and imports from the US) rose to 1.69m last year, which pushed GM into second place in the west European new car sales league behind the Volkswagen group, which includes Audi, Seat and Skoda. GM began production last year at two new assembly plants in eastern Germany and in Hungary.

It opened an engine plant in Hungary, where it has taken over as market leader. It expects to be in Europe with the launch this spring of a new generation small car to replace the Opel Corsa/Vauxhall Nova.

Financial Times, 20 January 1993

11.2 Examine the consequences of the decisions outlined in the article above for:
   (a) Volkswagen
   (b) SMH (SMH's new partner on this project is Mercedes-Benz – see Case study 10.3)
   (c) BMW (see tutorial exercise in Chapter 1)

11.3 (a) Examine the Sticky Sweets exercise on page 254. Suppose Sticky's competitors, Sharp, use a 4.5 per cent discount rate, while Spicy don't use any discount rate, but insist that each project should pay back within three and half years. What effect will these criteria have on the investment decisions and competitiveness of each company?
   (b) If Sticky, Spicy and Sharp moved to a WACC approach to set discount rates, what rate would they each set if the interest rate on bank debt were 10 per cent?

11.4 Firms in the UK tend to favour equity funding, while Japanese firms are debt funded (including public debt funds). What impact will this have on project appraisal, via the WACC-based discount rate approach?

11.5 Some management writers have referred to discounted cash flow techniques as an 'exercise in fantasy'. How could the statement be justified, and what are the implications for project appraisal and management generally.

Suggested solution to 11.1

Low volume

Year	Cash in £000	Cash out £000	Net £000	Discount 16%	DCF
0		2,000	(2,000)		(2,000)
1	4,000	3,400	600	0.86	516
2	5,000	4,200	800	0.74	592
3	6,000	5,000	1,000	0.64	640
4	5,000	4,200	800	0.55	440
					188

High volume

Year	Cash in £000	Cash out £000	Net £000	Discount 16%	DCF
0		4,000	(4,000)		(4,000)
1	6,400	5,000	1,400	0.86	1,204
2	7,200	5,550	1,650	0.74	1,221
3	6,400	5,000	1,400	0.64	896
4	5,600	4,450	1,150	0.55	825
					146

Clearly, the low volume output set-up has the highest NPV, and should be selected under NPV decision criterion. There are other factors to consider; a new product which is sensitive to price is not very encouraging; the vulnerability of the firm to market entry or a price drop of a substitute? How realistic is the marketing research, on which these projections are based? And will a small change in the discount rate make the answer different, or make both options unattractive?

12 An introduction to change management

Throughout the book we have raised the question of managing change within organizations. In this concluding chapter we examine change management in a little more detail. Many of the ideas discussed, such as motivation and organizational structure, do not lie exclusively in the change management domain, but are introduced in this context to show how they may become powerful forces to either resist or facilitate the management of change.

This chapter will begin by examining what it is that an organization may need to change. The remainder of the chapter will be structured around four stages identified as crucial in the change management process:

1 Pressure for change: normally caused by events inside or outside the organization that reduce the effectiveness of existing activity, present new threats or offer new opportunities.
2 A clear, shared vision: agreement and energy must be directed at formulating new missions and objectives (see Chapter 3), that are understood and accepted by everyone involved in the process of change.
3 The capacity to change: the resource base and attitudes of the staff must be consistent with the direction of change.
4 An achievable first step. Early successes give the process impetus, early failures cause false starts, recrimination and loss of capacity to change. It is in this last step that many engineers make a prime contribution, and this will be examined in more detail.

What has to change?

Since businesses and business environments are usually dynamic, change of some kind is entirely normal. The appointment or promotion of a new manager with a radically different personality, style or vision is likely to cause disruption and change. Entry into a new market segment may require new invoicing systems and credit facilities to be arranged, use of new technology on a production line will often necessitate design changes, re-tooling, retraining, recruitment and redundancy. In many ways, these changes concern the efficiency of the organization, that is, the degree of success it has in carrying out the things it does.

Although these changes may involve significant, and often traumatic changes to individuals and departments concerned, this chapter is concerned with wider scale changes of which the above may form a contributory part. Rather than efficiency, we wish to focus on effectiveness,

Figure 12.1 McKinsey's
Seven Ss

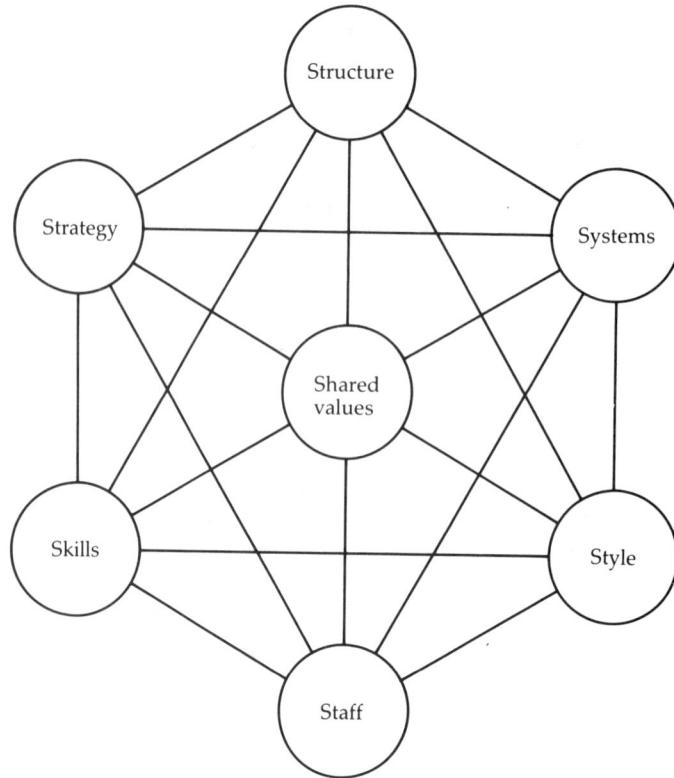

that is, the things the organization is trying to do. For example, a traditional bureaucratic organization may try to introduce a quality management programme by issuing a never ending stream of quality criteria forms,
documents and logbooks. The efficiency of the scheme could be measured
by whether or not the documents, logs and meetings are carried out and
the quality criteria met. However, the scheme is likely to be ineffective if it
remains a form filling exercise and does not engender the commitment to
quality that is necessary at grass roots level. Similarly, a stock control
system could be efficient if it enabled management to find out current stock
levels at any moment, but would be ineffective if it was clear that stocks-
out were still occurring.

Such 'effectiveness' problems tend to have much wider ranging effects
than efficiency problems, since they require the form to examine what it is
doing as well as how it is to be done. In this context (among others) it is
useful to consider the Seven S Framework used by McKinsey, in which a
specific organization can be defined and described into seven important
component parts, and the relationship between the parts can be analysed.
Conflict between the various parts can be a major barrier to organizational
effectiveness, and could be examined as part of the strategic audit discussed
in Chapter 3. Here we outline each part and suggest ways in which the
need to change any one part would be fundamental to the organization.

Broadly speaking, the Seven Ss can be sub-divided into hardware and
software. The hardware, the first three Ss discussed below, set out the

actions and activities which the organization is actually attempting to do. The software refers to people issues, that is, the way in which the hardware is made to work, or (perhaps), does not work. The final S, shared values, is the link between the two. Some commentators have argued that Western management thinking tends to be hardware orientated, particularly with regard to organizational structure, while Japanese thinking incorporates much more software activity.

Strategy

Strategy, in terms of the structuring of activities to meet objectives within the context of the organization's strengths and weaknesses and the environment in which it operates has been discussed at many points in the book. Any significant change in the business environment is likely to require major changes to strategy, with the attendant disruption that this will cause.

Structure

Structure refers to the way that an organization is structured. The Notts Knitting case on page 36 includes an organizational structure based on traditional managerial functions. Other structures might be based on product categories, geographical locations or, as is most likely, some mix of all these structures. Structure involves such issues as 'who reports to whom', 'who is responsible for what', 'who supervises what' etc. In stable markets and environments, it is possible to establish a strict hierarchy, in which each individual can identify their position accurately. In more volatile conditions, structures need to be able to respond quickly, by making appropriate decisions and acting upon them. Thus, inappropriate structure can be a break on change.

Systems

This refers to the way in which things get done. It would include such factors as production systems, investment appraisal systems, piece-rate working schemes etc. In stable environments, systems can remain unchanged, facilitating training and enable clear career paths to be mapped out. However, in dynamic environments, change may be forced upon an organization by competitors' action or a need to operate in a shorter time span, making decisions and acting faster. With quality circles, incremental change in systems should be continuous, regardless of the stability of the environment. Further, major changes in strategy and structure, and changes in systems allowed by, say, automation, may all require radical changes in systems. But strict job demarcation, custom and practice and inflexible management thinking towards systems may greatly restrict the scope for change management.

Style

Style refers to the way in which people, or more specifically management, goes about its activities. It incorporates important cultural values about what is regarded as important in terms of managerial time and attention, and the informal, unwritten rules that govern how an organization actually operates. In the paper exporting company referred to several times in this book, managers felt it was important to be restrained in their budgeting targets, and always remain within those budgets. Consequently, the firm's capital equipment became increasingly out-dated as managers tried to squeeze a few more years out of each item, and the R&D function was continually underfunded. Quite why managers believed this to be the best action was never fully understood, since the board of directors expressed a contrary view on several occasions.

It is important that style changes are not just urged on those required to change their behaviour, it must be clearly demonstrated that the old styles will no longer be tolerated. Of increasing importance in competitive markets is the way in which an organization treats its customers. For example, the difficulty experienced by privatized companies in changing the ways in which complaints are dealt with and enquiries are responded to is both a style and a systems issue.

Staff

People are the most important resource an organization has. The hard Ss are directed at making people do things, and there is a tangible element to staff, including such things as recruitment, training and appraisal systems. Intangible issues include morale and motivation. Both are important in ensuring that people are willing to do what is best for the organization as ably as possible. Quite simply, change management is impossible if the people concerned have been conditioned to resist change by, say, a reward system geared to other behaviour or falling morale.

Skills

Skills refer to expertise, or distinctive competencies that the organization has. It is these skills that enable a firm to establish a presence in the market-place. Ideally, these skills are used to best advantage by an appropriate strategy, and facilitated by the organization's structure and systems.

When there is a requirement to change the skills base of the company, these supporting Ss can continually push the company into applying and developing inappropriate skills. This was part of the problem at Notts Knitting, referred to in Chapter 2, where the need to develop a stronger marketing and sales function was inhibited by the firm's wish to depend on traditional design and production skills.

Shared values

Also called super-ordinate goals, this refers to the reason for the company's existence. This can be stated in the firm's formal mission, as discussed in Chapter 3, whereby all formal activity is directed at achieving the mission, and informal activity and commitment are orientated towards supporting those attempts. There is much evidence to suggest that a clear mission and sense of purpose throughout management are necessary conditions for effective change management.

However, it must be accepted that a mission is only suitable for a limited time period, and that the mission will change, hopefully subtly, at some point in future when it is less appropriate to the conditions in which the organization operates. At this point, the old mission and values may become powerful inhibitors, and yesterday's innovators and change managers become defenders of the old status quo (such entrenched people are sometimes referred to as risk averse people).

The kind of change we are examining in this chapter is that which involves effectiveness, and requires the organization to examine its activities and methods fundamentally. This inevitably requires major change in at least one of the Ss, and realignment of all the others. In order that these can be successful, the four-stage change management process is required. These four stages will now be discussed in turn.

1 Pressure for change

In this section we briefly outline the reasons why the need to change may emerge, and the barriers to acknowledging such change.

Internal pressures

Many writers have suggested that organizations go through periodic crises as the company evolves. Both the size and scale of the operations, and the technology used can thus bring about great pressure for change. Here, we outline Greiner's model of organizational evolution and revolution as an example of the kinds of pressures which can emerge.

A small entrepreneurial company (like Peter Smith's firm in Chapter 1) begins its life based on the skills and abilities of the owner, who is involved in every activity. If the firm is successful, the company's growth will cause problems, since the owner will not be able to lead every function simultaneously. Some owners do persist with this structure, but most cope by reorganizing the firm, usually by arranging the firm into departments and functions, with a leader to provide overall direction. However, further growth and development would imply that the organization and its functions become ever more complicated and specialist.

Leaders are unable to retain even an overview on complex and fast moving situations. Increasingly, the organization must find ways of

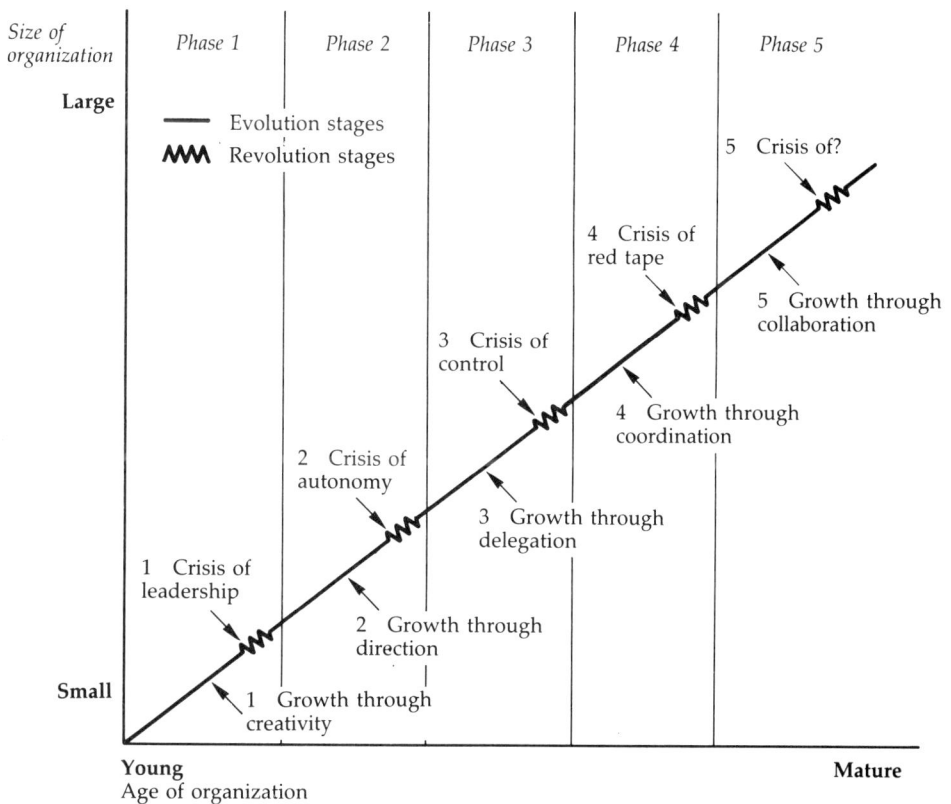

Figure 12.2
Greiner's model of organizationa evolution and revolution

269

delegating more power to those with knowledge of developing situations and appropriate operational skills. By this means, the organization continues to grow.

At this point, control issues are increasingly raised by senior management. Budgets, plans and reviews can all become means of control rather than expedition, and further growth may generate a bureaucratic system incapable of taking any initiatives. The approach required would seem to be based on co-operation, teamwork and creativity; and so systems driven by red-tape, protocol and procedures must either change or stagnate further.

All of the changes indicated above will require adaptation, and radical reappraisal of the seven Ss. And these crisis situations will develop in a successful firm in a completely stable business environment. Imagine then, the opportunity for disruption by a firm beset by uncertainty in a fiercely competitive market; and the inherent need for successful change management.

External pressure   The opportunities and threats identified in PEST and SWOT analysis, as discussed in Chapter 3, should identify areas in which the company needs to adapt and change. In particular, the competitive nature of global markets, and the opportunities and threats generated by technological innovation clearly make continual change the normal condition for many manufacturing sectors. As a reminder of the kinds of external pressure that can arise, reread the case study, Company Q, on page 235.

Barriers to change   Ideally, once opportunities and threats have been identified by a strategic audit, appropriate plans for change would be made and implemented. However, it is extremely common to find organizations that take no action in the face of a growing threat. There may be many reasons for this; including the possibility of strategists genuinely misreading a developing situation. More common, however, is the effect of filters and blocks to screen out messages which threaten the status quo. These would include:

Perceptual blocks   Where a problem is repeatedly mis-defined. For example, the problem of an ageing product range (i.e. where all a company's products are in the decline stage of the product life cycle), is really a strategic problem, requiring a radical rethink of the company's activities and direction. However, the company may see it as a sales problem, or a cost problem, and devise solutions inappropriate in the long term. Apricot Computers, for example, continued to believe that it was necessary to manufacture computers, even at a loss, in order to sell its maintenance and software services long after the computers had ceased to trigger service demand.

The danger of these blocks can be greatest when a refusal or incapacity to understand or acknowledge problems outside of a particular specialism can have serious effects. For example, accountants may decide to defer investment decisions or R&D projects for short-term financial reasons, as discussed in Chapter 11. However, such decisions have important marketing and engineering implications; for example, the company may lose the opportunity to deploy leapfrog technology, or may enter the market late, just as it reaches shakeout. These implications must be taken on board so that the full cost of the accounting decision in terms of missed opportunities can be assessed.

Cultural blocks

Organizations can create rules and myths which influence decision-making after any justification for them has long passed. For example, the major paper manufacturer referred to throughout the book was rife with such rituals. The company manufactured its own adhesives – oil-based adhesives at one location and vegetable-based gums at another. At some point, expertise had developed sufficiently for the production of effective hybrids, rendering the former distinction of sites obsolete. However, the firm decided to retain the distinction, because 'It had always been done like that', necessitating much internal traffic of part-finished glues from one site to the other. Only after several changes in management was the system changed to allow production of hybrids on both sites.

Examples of dangerous ritual beliefs include the conviction that a company's technology, product quality or after sales services are much superior to those of its competitors, and, most seriously, the invulnerability of the industry and company to competition from within the Single Market.

There are two particular difficulties with cultural blocks. First, no one may think to question them. To some extent, this is avoidable by using outside consultants and new recruits at a senior level. The second problem is that such blocks often take on the properties of a sacred cow, and it is very difficult to challenge the logic on which they are based since it is in the past (or mythological) rather than present. A good test for such thinking is to imagine that the proposed, changed state is in fact the present situation, and try to find arguments for changing to what is the current state. If no convincing arguments can be deployed, it is likely that the firm is suffering from a cultural block (try this by pretending that football or hockey is normally played on an artificial surface, and look at the gains and losses that would be made by transferring the game to grass).

There are other ways that pressure for change can be removed or dislocated from an organization's urgent strategic agenda. Many of these relate to the *style* element of the organization, such as who makes decisions, the pro-active attitude and responsiveness of the decision-makers to creative new ideas, the time horizons over which senior management think, their attitude to risk and their own perceptions of the business environment. Clearly, these blocks and filters are only capable of removing or diluting the perception of the need for change. Eventually, the real world will force itself on management, although, tragically, it may do so too late for effective reaction.

2 A clear shared vision

The McKinsey model places great importance on shared values, and clearly the other elements of the organization face great difficulties if they are not co-ordinated by common goals and values.

The implication of the management by objectives system described in Chapter 3 is that everyone knows the company's mission and objectives, sees their own role in the wider scenario and works to achieve individual and group objectives so that this shared vision is achieved. In reality, this is even more unlikely than it sounds, since there may be no such mission or objectives, and any that do exist may be rejected, ignored or sidelined by those entrusted to carry them out. In this section, we focus on means by which an individual may become included and motivated by the change management process. These possibilities are discussed under the headings of individual motivation, organizational factors and leadership factors.

Individual motivation

To individuals, change can be threatening and intimidating. New skills must be learned and previously well structured events and procedures will often be disrupted. Past contributions and improvements may well be discarded or made obsolete by changes. The effects of such changes as automation and computerization on the motivation of skilled staff are well known. Unless well managed, the changes are generally felt to be threatening and belittling. It follows that people who feel undermined and threatened by change are unlikely voluntarily to adopt and share the vision for the future that management wish to project.

Much has been written on the factors which motivate and demotivate individuals. Here, we look at the work of Hertzberg, who studied motivation and job satisfaction of engineers and accountants particularly, and whose work is widely used in industry. Broadly, he found that a class of factors, called motivation factors, were strongly linked to the content of a job, enjoyment of using professional skills and experience to achieve identifiable tasks, and recognition for such achievements, preferably through professional advancement. Lack of these things can lead to demotivation. A second class of factors, called hygiene factors, were more likely to be demotivators, and these tended to be derived from conditions of employment and working conditions, such as pay, administration, status, job security, relationships with superiors, juniors and equals etc. Hygiene factors were not, in themselves, likely to be good motivators, although could well demotivate when considered unsatisfactory.

Change and change management can be seen to have implications for both hygiene and motivation factors. Technological change may reduce motivation by deskilling and removing an individual from a task which was formerly enjoyable. Past contributions and achievements may well be wiped out. Skills formerly needed for advancement may become superfluous or redundant. Similarly, the individual may begin to think more about job security than advancement, particularly if the change has been initiated by a threat rather than an opportunity. Status, self esteem and the esteem bestowed by others may all begin to erode. Managerial errors and inconsistencies (inevitable during rapid change) and the general muddle of change management may be tolerated in other circumstances, but where motivation is poor, will be eagerly cited as another example of managerial incompetence, and a consequential backlash may deliberately undermine the company mission.

On the other hand, change can also be seen as an opportunity to develop new skills, enrich professional life and develop career advancement by adopting the changes. Hygiene factors can be reduced by renegotiation of conditions of service as job specifications change, relocation, and improved security if the changes are regarded as overdue and adding to the strength of the organization.

To some extent, the response will depend on the characteristics of the individual, such as age, personality, vested interests in the status quo, skills base and career plan. However, much can be done by managers if they bear in mind the effects that change can have on the individual. No matter how exciting, interesting and inevitable management might see the change, it is vital that hygiene factors are not allowed to dominate the experience of the individual.

Case study 12.1

Story of the software writers

A major overseas bank had a small London office, solely concerned with speculation on the London currency markets. They used a software package to maintain records of the deals struck, another package to keep accounts of money loaned to or borrowed from other banks and another routine to calculate the profit or loss made each day by each dealer, and translate these balances into the currency of head office. The company experienced difficulties with this software, the main one being that every programme had at least one major bug in it. No one was able to modify the software to remove these bugs, but several bright back-office operatives had devised spreadsheets which corrected the outputs. Without these corrected outputs, the dealers could not operate effectively. Once the skills of these operatives had been identified, they were continually called upon to write one-off programmes to automate cumbersome administrative and clerical routines. But in spite of their heroic efforts, by 1986 the volume of trading expanded so much that the major constraint on dealing was the time taken to update and process transaction details.

In early 1987 management purchased a new integrated suite of software, which replaced all the old systems and most of the bespoke software written by the back-office staff. The software was installed by a consultancy group who also delivered a first-rate training programme for users and an emergency call-out service; so that problems with the system were dealt with swiftly. The back-up operatives who had developed software no longer had such a role, and returned to their former clerical duties. However, promotions and advancement given earlier meant that they were paid more than their colleagues, but were not noticeably superior in their clerical roles.

In October 1987 came Black Monday, where complete chaos developed on all the financial markets. In the aftermath, the volume of foreign exchange trading was greatly reduced, and the bank had no need of such a large back office to process the dealers' transactions. The overpaid clerks were one of the first groups to be encouraged to leave the bank, when management looked around for cost savings in the back office.

Organizational
factors

In addition to the effect of change on the individual's established experience of work, motivation toward change and the degree to which the vision is shared will be influenced by the prevailing managerial style and the prevailing corporate culture. Some examples of the symptoms of a negative culture include:

- inability to get straight answers from people;
- lack of direction or accountability from senior management downwards;
- an obsessive avoidance of responsibility at all levels; buckpassing and disinterest, resulting in a lack of urgency;
- reliance on written memoranda to get others to act;
- a fall back on job demarcation even among office administration staff.

The many books written on 'culture change' will testify that the process of change in workforce attitudes can take more than three years (and often much more) and if it is accelerated by a 'force-feeding' process, can result in a more dangerous situation than that of outright strikes: namely that of passive resistance of the sort that is the most difficult to identify – a negative culture with covert undertones that are almost impossible to define. In this case, owners may be excused for feeling that the only solution is to start again, carrying a nucleus of core skills necessary to develop on a 'green field' site in a new environment.

The process of change management itself can easily generate a negative culture, even where there was no such attitude before. Some elements of the process for reducing these tendencies are discussed in the section on coping with change.

The degree to which a vision of the company is likely to be shared by everyone is also influenced by the managerial style of key people involved. These managerial styles have been derived and described in a wide variety of ways by different writers; here we focus on three styles referred to as X, Y and Z.

Theory X

In many senses this is traditional, nineteenth-century management thinking, where managers plan and organize the workforce, and monitor and control activity to check that their orders are carried out. It tends to rely on the assumption that workers are lazy or unwilling to take initiatives, dislike work and effort and will only operate to achieve the greatest reward, in terms of pay, and minimize pain, in terms of retribution, dismissal etc.

Theory Y

This theory is perhaps more enlightened in that it assumes that work is as natural as rest and play, and that negative attitudes arise largely from bad experiences rather than intrinsic dislike. Under these assumptions, management activity is directed at maximizing the contribution individuals are able to give by allowing self discipline rather than supervision, problem solving rather than instruction, and ensuring that these activities are directed toward the organization's best interests. The theory also implies that negative attitudes are a symptom of bad management, rather than the natural condition of the work-force.

Theories X and Y have been discussed interminably throughout the West. It is clear that both can be made to work by able managers, but in the context of change management, both have their strengths and weaknesses. Theory X managers may be able to force change through very quickly, a major advantage in a fast moving situation. However, this will not generate a shared vision or encourage commitment to the organizational mission. Indeed, this cannot be so since the theory assumes that the interests of most of the workforce will be intrinsically opposite, and implicitly hostile, to the mission of the company. Theory Y, on the other hand, suggests that managers would seek to involve staff in the change management process by delegating and discussing objectives and methods so that people are able to implement part of the process for themselves. This attitude, if applied fairly and well, is more likely to encourage a shared vision.

Theory Z

This management style has been developed furthest in Japan, and can be described in a variety of bewildering ways. For our purposes here, we would say that the management style is associated with many employment practices which reduce hygiene factors, and foster motivation factors. Hygiene factors are minimized by such things as guaranteed job security and promotion from within, and an emphasis on shared responsibility for decision taking and implementation. Moreover, career paths are horizontal as well as vertical; meaning that promotion and re-assignment are likely to transfer an individual between managerial functions as well as up them.

The consequence is that individuals are more committed to the goals of the organization, since they would have had some part in setting these goals. People are thus empowered to bring about their own changes, and management is seen as a guide and a resource. Plan ownership is thereby transferred to its lowest level, and therefore becomes operational far more effectively. Further, such a style not only responds to change effectively, but is actually a major instigator of pro-active innovation through quality circles (see Chapter 5). Many would therefore argue that Theory Z styles are most likely to generate a long-term strategic thrust, articulated through a clear, shared vision. The downside is that the consultation processes may be much longer than with Theories X and Y, although the delay may easily be made up by a more rapid implementation once agreement has been reached.

Leadership factors

We have looked at the difficulties of winning an individual's commitment and energy to a vision by virtue of the changes on his or her actual job, and the impact of participation in that decision through managerial style. We conclude this section by looking at the potential for leaders to carry their people with them toward the new vision or mission. We shall do this by assessing how common 'types' of leaders may motivate and inspire their people.

Charismatic leaders

Often able to inspire others, either by charisma or some other persuasive style, to reach for some compelling vision. Clearly a major asset for change management, providing they are leading their people towards company, rather than personal goals.

Crisis leaders

Leaders who naturally arise from circumstances, and disappear when those circumstances change. Such leaders are crucial in project management teams and task forces, to use their expertise and skill to push through such change as is necessary, but then fade when power can be transferred further down the line.

Contingency leaders

These are natural leaders, who take on board both the tasks necessary to carry out an operation, such as a systems change, but also the needs of the group and the needs of the individuals concerned. Such a leader may have limited expertise in some areas, but can make an ideal group leader.

The contingency manager will adjust behaviour and style, depending on the situation in which he or she is operating. Providing such leaders are committed to change, the overall approach can do much to generate sympathy and commitment.

Positional leaders

Such leaders depend upon their formal role and position in the hierarchy, rather than personal qualities or abilities. Functional managers, including engineers, can often become leaders of this type, as advancement takes them further from their training and expertise. In few cases do promoted leaders become contingency leaders automatically following their promotion. It can take considerable training and management development to help positional leaders become effective change managers rather than risk averse people.

Succession leaders

Such leaders arise for reasons largely unconnected with leadership ability, or any other ability for that matter. In business, this may be the normal case in family-dominated firms, where successive family members occupy most senior positions, perhaps for generations. Similarly, many managers attempt to groom a successor, chosen for reasons which may have little to do with outstanding managerial competence.

Such leaders may be tempted to become risk averse people, and would not wish visions to be shared, since their power lies in the status quo. However, this is not inevitably the case, and such leaders can become any of the other types listed above.

Summary

Much can be done to reduce the possibilities and effects of unclear, fragmented missions and objectives. In Chapter 2 we looked at the SMART system of objective-setting. Used in conjunction with employee appraisal, also based on MBO, the role of each individual can be steered appropriately, and individual dissent managed in a way that does not threaten the change management process. However, this must be backed up with due regard to the hygiene and motivation factors which will affect the individual at work, the extent to which the managerial style will communicate the vision and involve staff in its implementation, and the leadership qualities which will energize the whole thrust of change. Failure to establish a shared vision will mean that the best-made strategic plan will never cascade throughout the organization, and co-ordinated, fundamental change will not endure.

Capacity for change

Capacity for change can be examined on two levels. First, capacity should be assessed as part of the strategic audit, where the quantity and quality of financial, managerial and technical resources are assessed against the resource implications of the strategic change proposed. It is self evident that a change management programme which requires resources beyond those currently available to the organization, cannot be pushed through to completion. Company Q, discussed on page 235 has fallen into this difficulty, and unless extra resources can be generated speedily, faces an uncertain future. These issues have been raised several times in the book, in particular in Chapter 9 and 10, and will not be reviewed here.

The second level on which capacity should be examined is the question of organizational impediments to change. This includes hardware issues, such as organizational structure, and software issues, such as learning and coping with change. These are discussed below.

Organizational structure and change

Organizational structure refers to the ordering of individuals into departments, sections and groups in order that the business of the organization can be carried out.

An extremely common structure is that designed around managerial functions, so that accountants are all in one department, engineers in another and production in a third etc., as shown in Figure 12.3.

The grouping of such specialists has obvious advantages, and chains of command and accountability can be made quite clear. Similar hierarchies can be developed along product lines, where all activities necessary for the design, production and marketing of products are grouped individually. Further variations for hiearchies based on key customers, customer groups and geographical location are not uncommon. Such a structure is referred to as mechanistic, meaning that loyalty, orders and interactions are passed up and down the hierarchy, and tasks, duties, roles and customs are clearly demarcated and often sacrosanct.

The benefits of a hierarchy are self evident, particularly when most of the work of the organization is routine, and capable of being sub-divided, so that strategy and complex issues remain at the top of the organization, and simple and operational matters can be passed down to those who will supervise and carry out the work itself.

However, many studies have shown the difficulties that a traditional structure has with change management. This must be so, since the hierarchy will depend upon the very rules, procedures and customers which may need to

Figure 12.3 The hierarchical organization

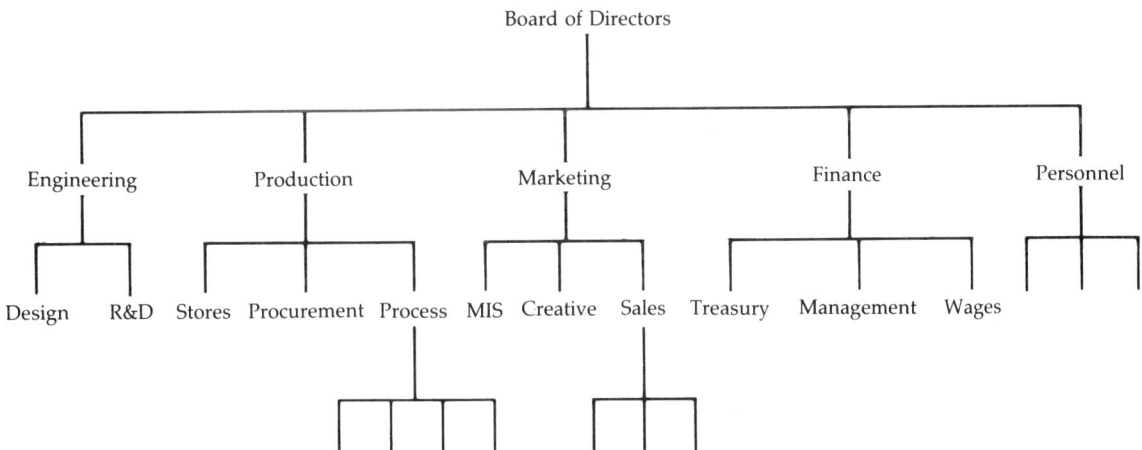

Board of Directors

Engineering   Production   Marketing   Finance   Personnel

Design   R&D   Stores   Procurement   Process   MIS   Creative   Sales   Treasury   Management   Wages

be changed. But perhaps the major weakness is that the multi-disciplinary, multi-tasked project group, so necessary for effective change management, does not arise naturally from such structures, and so structures generally lag behind important changes in operations or the environment. New managers recruited into such a hierarchy may often wonder how the structure and attendant systems might ever have worked. When faced with the 'we've always done it this way' syndrome, they have only two choices:

1 They either toe the line until they get their feet under the table, or
2 They are encouraged to be radical in approach and are supported (often covertly) by superiors who appointed them deliberately to break the barriers of convention, so that a fresh approach is steamrollered through a culture resistant to change.

Item 2 seldom works without the implicit cooperation of the existing workforce, even when carried out by management consultants, because the timescale expected by the owners of a business to achieve a result is usually too short for meaningful attitude- or culture-change to take place.

At the more flexible end of the scale, structures with more lateral flows of communication and task division are often referred to as organic structures or alternatively networks. An example of a hierarchy which adopts project teams for specific problems is shown in Figure 12.4.

In such an organization, routine events are still dealt with in a traditional hierarchy, but, of more significance, are the matrix organizations running from left to right. These teams are multi-functional, designed to focus the correct mix of skills at a specified problem. Of course, the structure can be undermined if hierarchical considerations still apply, and team members are regarded as representative of a function rather than contributing experts. But when the system works properly, it normally shows behaviour much more conducive to change management, such as the horizontal flow of information and expertise rather than directives, a focus on issues and problems rather than organizational politics, a commitment to the project as ownership is transferred to team members and so on. The vital contribution which can be made by project teams is discussed in the concluding section of this chapter.

Capacity for change can be restricted by an organizational structure designed to operate under conditions of stability and efficiency. However, changing the formal structure will not in itself bring about fundamental change, as many companies have found after years of 'restructuring'. Changing the titles on office doors and placing bureaucrats in teams still leave the other six Ss unchanged.

Coping and learning

The people in the organization must have time and support to cope with changes and learn the new systems and styles necessary, even if they are committed to such change.

Here we outline the coping cycle proposed by Carnell (see reference in the 'Further reading' section towards the end of this chapter), which indicates the kind of stages that changes manager should expect staff to go through. Capacity depends upon the speed at which people can be encouraged through the cycle.

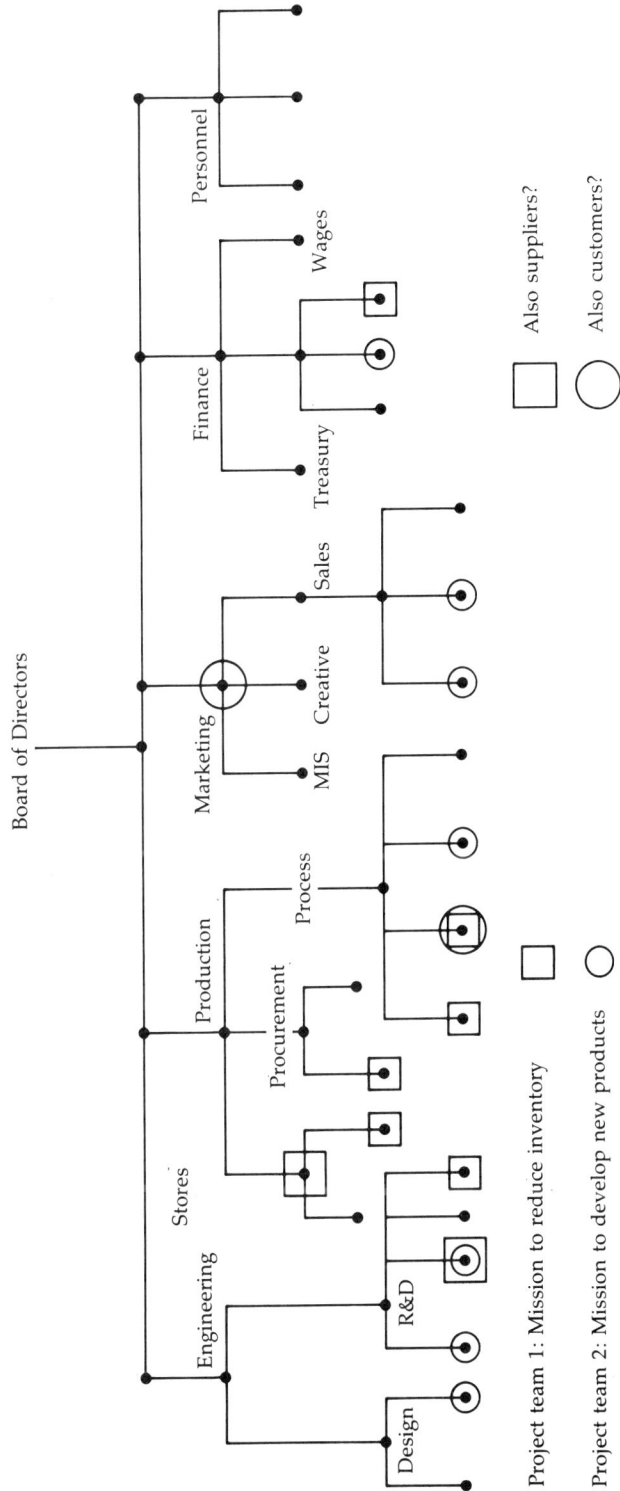

Figure 12.4 Composition of project teams in the hierarchy

Board of Directors

Personnel

Finance — Wages, Treasury

Marketing — Sales, Creative, MIS

Production — Process, Procurement

Engineering — Stores, R&D, Design

Also suppliers?

Also customers?

Project team 1: Mission to reduce inventory

Project team 2: Mission to develop new products

| *Denial* | The initial reaction to change proposals is almost invariably negative, for reasons outlined in the section on shared vision. In this stage, management must work to change attitudes, a premature introduction of new systems and structures would compound the initial resistance. Shared values and a clear vision will only arise as problems are discussed and shared. |

Defence

Once negative attitudes towards change in general have been overcome, it is necessary to deal with specifics. In this stage, individuals may seek to place the balance of disruption elsewhere by showing how past achievements are consistent with the required change in behaviour, or by implying that the required changes do not relate to their particular area or function. This defensive attitude may be short lived and ritualistic, but generally speaking, the process of instilling plan ownership and problem solving can begin at this stage.

Discarding

At this stage, much of the resistance to the change is lost as staff become committed, to some degree at least, to making the changes work. At this stage, management must be most supportive, since changing from one system or process to another is bound to lead to mistakes and stresses. Genuine mistakes and errors should be tolerated, with training and counselling being generally more effective than disciplinary action or obtrusive monitoring.

For example, a major UK producer of photocopying equipment computerized part of its clerical function in the late 1970s. The changes were traumatic, involving loss of status and redundancy for some clerks. However, mistakes in completing the actual computer input slips were never criticized or reported to management. When the data input operatives detected an error, it was taken to the clerk in person and the correct procedure re-explained. The disruptive effects of the redundancies and changes in job specification lasted a long time and caused much bitterness, and so could have been better managed, but the computer system ran efficiently in a remarkably short time.

Adaptation

In most cases, adoption of the change does not end the process. New systems and procedures inevitably go wrong. If the earlier stages have been managed correctly, staff will begin to accept ownership of the new systems by adapting them through improvements and repair. With good management, this adaptation and improvement behaviour could be continuous. Here, managerial support is best in the form of rewarding (which may include sincere thanks!) and using expertise to support those making innovations.

Internalization

The final stage is where the new, changed system is now regarded as normal accepted practice, subject to whimsical complaints from a few remaining risk averse people. The change management process is now complete; until the next change in the business environment initiates the process again.

Managing people in this way will not come naturally to many managers, particularly those who are naturally Theory X managers or in organizations where Theory X is the predominant style. This greatly reduces the capacity for effective change. Further, the cycle may require considerable time to develop; perhaps years if the changes are particularly dramatic. However, the problems faced in the business environment may not allow the luxury of a protracted transformation period.

Summary

In this section we have looked at a firm's capacity to change, in terms of the organization's resource base in relation to the changes proposed, the ability of the organizational structure to adapt and to foster change, and manager's ability to introduce change in a fashion which tolerates and overcomes natural resistance. Failure to develop a capacity to change prevents the programme reaching full implementation, and may well obstruct other changes as staff can become cynical over management's inability to deliver actionable programmes.

Actionable first steps

The first steps of any change management programme must be seen to be successful, and create a dynamic climate of improvement. Thus, the first steps of a programme should be selected on the basis of the likely chances of success and the speed with which obstacles can be overcome. Project teams are central to the success of this stage and are discussed in this section, although their importance to change management has been identified throughout the book, and this chapter in particular.

The role of project management

This chapter reports on a well-tried process of managing change on an evolutionary basis – and, if carried out correctly, it can achieve the whole-hearted support of the workforce, because they in fact helped to create and then implement the plan for change. The hidden agenda of project management is that the workforce ends up 'owning the plan'. It is carried out best by existing management and workforce.

Plan ownership develops a momentum of its own, whereby the commitment and enthusiasm among the workforce, as well as at all levels of management results in the whole company facing in one direction; this is the recipe for making 'impossible' missions achievable.

The difficulty is to get from 'talking shop' to meaningful action, particularly those important first steps. Setting up a project team involves finding the small percentage of pace setters in an organization; no easy task, particularly when a marriage of skills is equally important for forging teams made of 'the right stuff'. The objectives entrusted to the project team have to be concise, measurable and unambiguous; and management must encourage trust by assuring the team that when it comes to spending money, the funds are really there to implement the change, be it a new product development project, a factory move, a market penetration programme abroad or a combination of all three. It is this last point that distinguishes a dynamic project team from a flaccid committee. The idea is not to report back to some other body, or produce a report to be debated interminably, it is to entrust the project team with finding and implementing solutions in a much more direct fashion.

Project management and boundary management

Boundary management is the process of managing complex interfaces between skills and functions required to carry out a task. It is a problem at an individual level, i.e. how does a marketer or an engineer learn to cope with the complicated people and budgeting issues which may follow a major promotion? Similarly, it is a problem at an organizational level, that is in what way does the organization assemble the wide variety of skills and experience necessary to crack a particular problem?

Modern management is a complex compromise, where the matrix structures prevalent in project teams are superimposed on classic hierarchical structures. The latter will always be needed to ensure that the routines of everyday business are carried out smoothly. The problem is that the very stability of hierarchical structures tends to atrophy attitudes into sticking with what is known to work, until a major external change threatens the very existence of the organization. At this stage it is necessary to build project teams with the implicit cross-boundary credibility and clout to get commitment to change and the vital cooperation from the workforce to avoid total chaos when it most matters – at the early stages – when simple mistakes can have a devastating effect on morale.

As an analogy, project management can be likened to providing the oil for lubricating the parts of successful boundary management. It is crucial that the 'advance scouting team' entrusted with starting the change process leads by example: and the first challenge is to demonstrate that they can be effective together. The right blend of personalities, balance of various psychological outlooks and choice of the 'skills base mix' are all critical, because they will provide the subsequent 'new blueprint' to emulate throughout the organization.

Examples of this are many:

- The Japanese can manage continual change successfully, even while ensuring consensus across a wide spectrum of the organization. They do not seem to need the pilot teams to forge ahead of the rest of the organization. The whole workforce is adaptable to change.
- The now famous Volvo experiment in Sweden, where multi-disciplinary assembly teams replaced the single-disciplined assembly line mentality, forging the early ideas of quality circles and 'group accountability' for excellence, largely eliminating the need for inspectors. This goes back to the 1960s.
- The project management companies who support the oil industry, such as Davy McKee, Eurotunnel and Tarmac in construction, continually practise their own brands of project management, handling multi-million pound contracts, each of a unique specialist nature. The project team members must be engineers, procurement specialists and financiers all rolled into one – namely, they are rounded business people as well as possessing a functional specialism such as IT, fluid power engineering or marine engineering. They only get one chance to get it right, so that even at the planning stage they must have a good idea of the cost of the project – materials, labour, inflation over a possible three-year period, subcontracting costs for specialist services etc.

At this stage, it is important to go right back to Figure 10.7 in Chapter 10 on page 246, to recap the role that project management could play in any profit-making as well as non-profit making organization.

The axiomatic principle is to work at 'plan ownership' down the line. This is fine in theory, but difficult to achieve in practice, if the workforce is not involved in the early stages of conceptual planning.

The self-defeating dichotomy here is that to involve the practitioners at the early stages produces 'To do' lists of everyday issues (we've always done it this way syndrome). Unless management have enough finesse, and the trust of the workforce, they are unlikely to get even middle management to contribute at a strategic level; after all it is something middle managers have traditionally been excluded from.

The types of company who are most likely to consider project management voluntarily are those who have had to face change the hard way, and have come through to tell the story. The very nature of the problems has produced the right leaders for the day, who have helped bring down traditional boundaries so that workable project teams have been formed, either informally or by formal appointment.

Project management is also instrumental in easing the individual aspects of boundary management, in that it fosters a sense of the need to retrain/train in non-core skills – namely, engineers in finance and marketing, and managers in the skills of boundary management and the fostering of durable and resilient cultures by developing a clear concise corporate *raison d'être* which the whole workforce has taken apart in building.

This is clearly consistent with the procedure outlined in Chapter 3:

- Where are we now?
- Where should we be?
- How are we going to get there?
- What shall we do?
- How will we know when we've got there?

If these steps are not taken in this order, the resulting plan may be little more than an annual budget plan rather than a far-reaching strategic plan which encompasses and ensures the ongoing health and real growth of the organization. 'Growth' can range from profit increase with a drop in turnover, to controlled temporary profit decline in favour of essential market share increase, involving investment in product design and distribution systems before rewards can be reaped in terms of incremental turnover increase, productivity gains and the establishment of brand loyalty in, say, a new country or continent. Growth in non-profit organizations can include, for instance, setting ambitious productivity increases, achieving new levels of service and environmental responsibilities.

Case study 12.2: Revitalizing a company

The road paving machinery company

In 1984, a manufacturer of construction equipment, renowned worldwide as an innovative multinational force in road paving technology, recognized that their present range of products had lost reputation as technology leaders. Symptoms included the need to sell at cost price

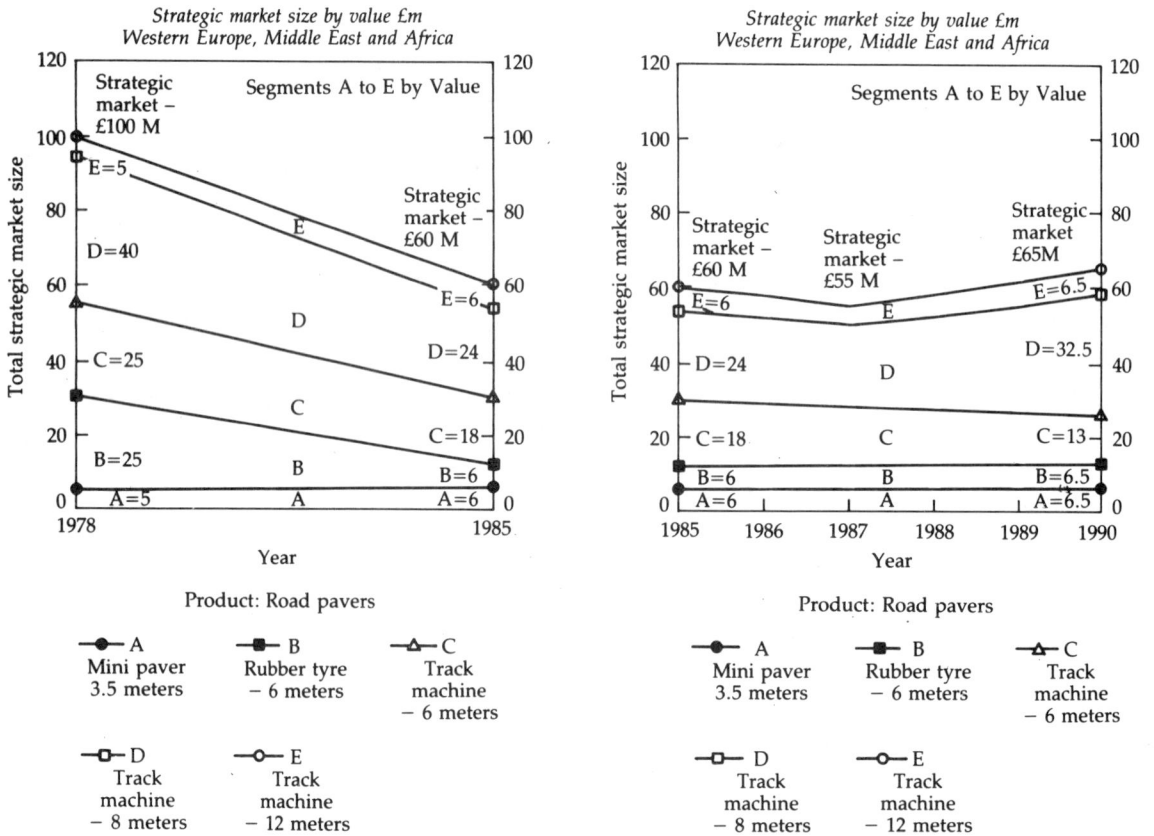

Figure 12.5 Strategic market size: actual trend over previous seven years and forecast five years forward

to compete with recent established European competition, a plethora of optional features to be included in the standard price, and extensive grumbling from the company's own salesforce and its international dealer network that there remained no competitive edge to sell. The company was losing the 'specification battles' that are prevalent in capital goods markets, particularly where public tendering with life cycle planning of five to seven years was involved.

The business was operating at a loss, but still retained a formidable reputation with the financiers who supported this family-run multinational. The markets in which it operated became increasingly turbulent, causing poor forecasting of even the short-term order book (three months ahead). The problems were further compounded by the severe and sustained recession, in which the sales of road making capital goods had dropped by 40 per cent between 1980 and 1984. The company believed that the consortia of bankers who supported the company were fast losing patience. The feeling of impending crisis triggered a strategic audit, in the hope that a rescue plan could be devised. In particular, the banking consortia wanted to see far-reaching missions, complete with detailed action plans, and projections of

Strategic market size by value £m
with company market share
Western Europe, Middle East and Africa

Segments A to E by Value

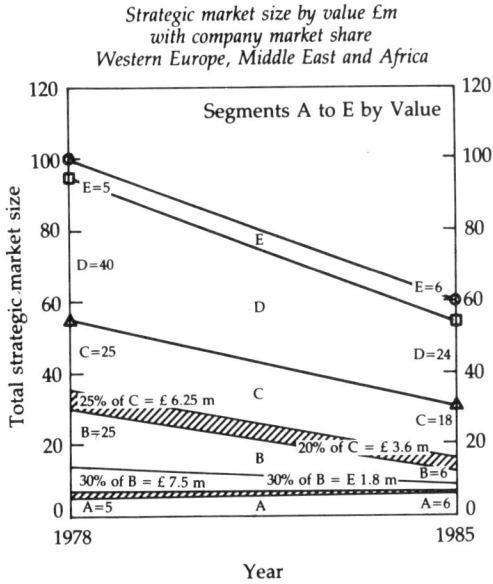

Year
1978 — 1985

Product: Road pavers

- ●—A Mini paver 3.5 meters
- ■—B Rubber tyre − 6 meters
- △—C Track machine − 6 meters
- □—D Track machine − 8 meters
- ○—E Track machine − 12 meters

Strategic market size by value £m
Western Europe, Middle East and Africa

Segments A to E by Value

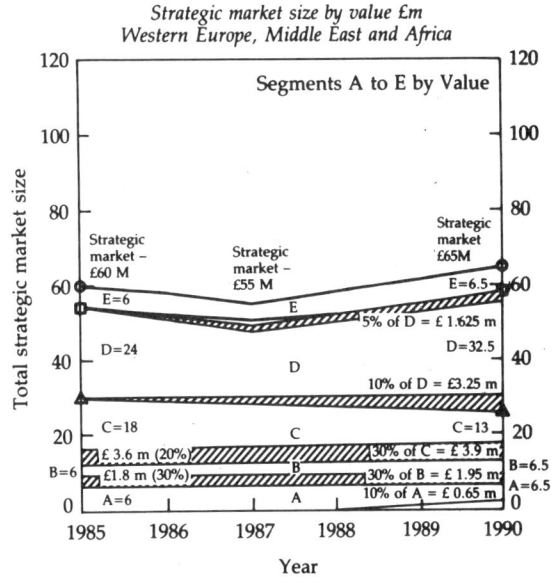

Year
1985 1986 1987 1988 1989 1990

Product: Road pavers

- ●—A Mini paver 3.5 meters
- ■—B Rubber tyre − 6 meters
- ▲—C Track machine − 6 meters
- □—D Track machine − 8 meters
- ○—E Track machine − 12 meters

Figure 12.6 Strategic market share: actual shares of served markets with existing range. Future projected shares with new range of products

the financial consequences (presented on a weekly cashflow basis) of these action plans.

The major problem identified by the audit was that of poor market segmentation. The first step to correct this was to identify key segments by product group, as shown in Figure 12.5. Once these segments had been identified, market size measured and trends projected, it became possible to build up a realistic picture of the market. Finally, the firm's existing products were plotted onto two graphs (see Figure 12.6). A major part of the problem immediately becomes clear. It is apparent that the portfolio of products offered by the company left several major sectors unserved. In particular, the failure to position a product in the largest segment, that of 8-metre track-laying vehicles, was a major omission. Therefore, any strategic mission had to be based on placing a market leading product in this segment; which was the only sector expected to grow.

Further auditing showed that launching a new product range would not be enough, the firm was not particularly competitive, as costs were too high. Action to improve the competitive position consisted of the following:

285

- Sale of factory which was too big; hence providing the finance to pay off the overdraft which had reached its ceiling of several million pounds sterling. It would then be realistic to rent/lease smaller premises suitable only for the assembly process of the products.
- Investment in assembly methods to ensure global competitiveness by minimizing labour content.
- A radical reappraisal of materials management, including complete subcontracting of all weldments, major fabrications and many component sub-assemblies. What was unwittingly forced on the company was the makings of a 'just in time' system dictated by a critical cash shortage.
- A need to write off a third of the company's stock as basically worthless scrap: much of it had an age of five years or more, and was in any case obsolete through uncontrolled design changes.
- Radical redevelopment of the products to achieve (outlined in Chapter 10, Figure 10.7):
  - a 30 per cent saving cost of sales;
  - a balanced range of products to reflect significant shifts in market sub-segments;
  - a need for the highest possible component commonality across the new range of models; to give some competitive edge against West German, British, US and Scandinavian products of the highest quality and proven reputation in service;
  - a sensible range of optional features for the international dealers to offer maximum differentiation according to national preferences, for example, lower noise levels in the French market, TUV brakes in West Germany and operator cabs, complete with computerized steering controls (heated in winter and air conditioned in summer) for Sweden.

Above all there was a continued nervousness at the poor understanding of the market dynamics at play, shown up in persistently over-optimistic forecasting. A classic symptom allowing too much responsibility for marketing thrust to international dealers, which had proved almost fatal, because dealers were there to make money first and foremost, and focused on the franchises that helped them do so. Selling and inventory control efforts were prime, rather than extensive attention to gathering marketing information for strategy discussions. PMC's products went to the bottom of their priority lists, as dealers strived desperately to find money-spinning alternatives.

The fully costed and operationalized details of the plan were shown to the consortia of bankers, who were anxious to see a credible rescue package put together. Leaving nothing to chance, they insisted on a 'management by objectives' strategy which they monitored monthly, and dictated seemingly impossible deadlines for:

1 A factory move: four months allowed after sale of facilities.
2 A product development plan involving additional funding of £¾ million, providing the first paver prototype was launched in half the best time period achieved by the company's parent organization in the USA (nine months rather than eighteen months).
3 Manufacturing engineers and prototype buyers were to be part of a team entrusted with reducing the number of line items per machine by a half: from 3,500 to 1,700, while the designers had to design for over 40 per cent component commonality across the range.
4 An immediate obsolete inventory clearance plan to arrive at a believable stock valuation on the balance sheet.
5 A resource reorganization plan that involved the recruitment of CAD literate engineers capable of working and leading in project management mode, including taking competitors' machinery apart on site to generate a lasting competitive edge in product features and benefits and enable a 'Me Too Plus' strategy to be followed (see Chapter 10, Figure 10.7 refers).
6 An investment plan to automate the assembly process with much greater use of capital machinery, and automating the delivery of components straight to ten assembly points on the line, so that there was minimal warehousing of components; just safety buffer stocks at the point of use allowing for a week's lead-time between orders (this was not quite the JIT philosophy of no safety stock, but was getting close).

Although this plan was tight, and depended on some highly optimistic market projections, the major strategic targets were met.

Chapter summary

In the concluding chapter we have examined the change management process, looking at the four key stages of change management and showing why project management is a vital ingredient.

Further reading

A great deal of human and organizational behaviour theory is well represented in the relevant sections of *Management, Theory and Practice*, by G.A. Cole (DPP, 1990). We would also heartily recommend *Understanding Organisations* by C. Handy (Penguin, 1993). The readings in *The Strategy Process* by H. Mintzberg and J. Quinn (Prentice Hall, 1991) are helpful for advanced reading.

Books dealing explicitly with change management include *Managing Change in Organisations* by C.A. Carnell (Prentice Hall, 1990) which combines theory with practical suggestions, and *Thriving on Chaos* by Tom Peters (Knopf, 1987) which is more anecdotal in many places, but a good read anyway. More serious studies include those by R. Moss Kanter, including *Change Masters* (Allen and Unwin, 1985) and *When Giants Learn to Dance* (Simon and Schuster, 1989). It is always worth reading the accounts of successful industrialists, such as J. Harvey-Jones's *Making it Happen* (Fontana, 1989).

A contrary view to many of the arguments presented in this book, and much of the suggested reading lists, is presented in *Managing Chaos* by R. Stacey (Kogan Page, 1992) and *The Chaos Frontier* (Butterworth-Heinemann, 1991) by the same author, and both are strongly recommended.

There is an increasing number of helpful texts on project management, including *Project Team, the Human Factor* by O. Kharbanda and E. Stallworthy (Blackwell, 1990).

Tutorial exercises

12.1 Consider the case, 'Story of a Software Writer' on page 273. Examine the hygiene and motivation factors which would apply to the software writing clerks. If the clerical section were run by a Theory X style manager, how would the clerks be treated? Would a Theory Y or Z style manager operate differently? How should the case conclude, according to each theory?

12.2 Examine the market segmentation data given in the case study 'Road Paving Machinery'. Which market segments would you abandon, and which would you focus upon?

12.3 With the information in the case, and your analysis from 2. above, develop the fourth stage of planning: that of action/implementation plans to this situation. Suggest areas best tackled by project teams, and indicate the type of specialists required. How would senior management fit in with these new project teams?

Index